Advanced Ta-Based Diffusion Barriers for Cu Interconnects

ADVANCED TA-BASED DIFFUSION BARRIERS FOR CU INTERCONNECTS

RENÉ HÜBNER

Nova Science Publishers, Inc.
New York

Copyright © 2009 by Nova Science Publishers, Inc.

All rights reserved. No part of this book may be reproduced, stored in a retrieval system or transmitted in any form or by any means: electronic, electrostatic, magnetic, tape, mechanical photocopying, recording or otherwise without the written permission of the Publisher.

For permission to use material from this book please contact us:
Telephone 631-231-7269; Fax 631-231-8175
Web Site: http://www.novapublishers.com

NOTICE TO THE READER

The Publisher has taken reasonable care in the preparation of this book, but makes no expressed or implied warranty of any kind and assumes no responsibility for any errors or omissions. No liability is assumed for incidental or consequential damages in connection with or arising out of information contained in this book. The Publisher shall not be liable for any special, consequential, or exemplary damages resulting, in whole or in part, from the readers' use of, or reliance upon, this material. Any parts of this book based on government reports are so indicated and copyright is claimed for those parts to the extent applicable to compilations of such works.

Independent verification should be sought for any data, advice or recommendations contained in this book. In addition, no responsibility is assumed by the publisher for any injury and/or damage to persons or property arising from any methods, products, instructions, ideas or otherwise contained in this publication.

This publication is designed to provide accurate and authoritative information with regard to the subject matter covered herein. It is sold with the clear understanding that the Publisher is not engaged in rendering legal or any other professional services. If legal or any other expert assistance is required, the services of a competent person should be sought. FROM A DECLARATION OF PARTICIPANTS JOINTLY ADOPTED BY A COMMITTEE OF THE AMERICAN BAR ASSOCIATION AND A COMMITTEE OF PUBLISHERS.

LIBRARY OF CONGRESS CATALOGING-IN-PUBLICATION DATA

Hubner, Rene.
  Advanced Ta-based diffusion barriers for Cu interconnects / Rene Hubner.
    p. cm.
  ISBN 978-1-60456-451-8 (hardcover)
  1. Interconnects (Integrated circuit technology) 2. Integrated circuits. 3. Electrodiffusion. I. Title.
  TK7874.53H83 2008
  621.3815--dc22
                                    2008008583

Published by Nova Science Publishers, Inc. ✢ New York

CONTENTS

Contents		5
Preface		7
Chapter 1	Introduction	9
Chapter 2	Experimental Details	31
Chapter 3	Microstructure and Functional Properties of As-Deposited Ta-Based Diffusion Barriers	37
Chapter 4	Thermal Stability of Ta-Based Diffusion Barriers on Sio_2	43
Chapter 5	Trace-Analytical Techniques for a Sensitive Proof of Cu Diffusion	63
Chapter 6	Conclusion	77
Acknowledgments		79
References		81
Index		95

Preface

During the last few years, copper has become the standard metallization material for on-chip interconnects in high-performance microprocessors. Compared to the previously used aluminum, copper shows not only a lower resistivity, but also significantly improved electromigration resistance. Copper ions, however, are very mobile in silicon and many dielectric materials under electrical and thermal bias. Thus, barrier layers are needed to prevent Cu diffusion into the insulating layers surrounding the metallic interconnects. Since Ta-based compounds are characterized by a high thermal stability, pure Ta films or layer stacks consisting of Ta and TaN are used for such barriers. The continuous scaling down of the interconnect dimensions and, therefore, the essential decrease in the barrier layer thickness coupled with the replacement of silicon oxide by advanced low-k dielectrics demand further improvements of the diffusion barrier performance. It is the aim of this study to carry out microstructure and functional property investigations for advanced, high-performance Ta-based diffusion barriers (Ta-TaN layer stacks and Ta-Si-N single layers) before and after annealing to compare their thermal stabilities and to probe the corresponding failure mechanisms. For the Ta-TaN barriers, these studies are undertaken for a range of layer sequences, while for the Ta-Si-N barriers a variety of films with different chemical compositions are analyzed. One precondition for a diffusion barrier film is the chemical stability with the interfacing dielectric. To examine this stability, annealing experiments are performed for uncapped barriers deposited onto silicon oxide. Upon addition of a Cu metallization layer, it is shown that the combined application of X-ray scattering methods (X-ray reflectometry, glancing angle X-ray diffraction), spectroscopic techniques (glow discharge optical emission spectroscopy, Auger electron spectroscopy), and microscopic analyses (transmission electron microscopy, scanning electron microscopy) is suitable to investigate subsequent microstructure changes during annealing. In particular, the phase formation behavior and the interdiffusion of tantalum, silicon, and nitrogen are characterized. The sensitive proof of Cu diffusion through the barrier films is carried out based on two different approaches. Firstly, trace-analytical techniques (atomic absorption spectrometry, secondary ion mass spectroscopy) are applied at barriers prepared between copper and silicon oxide. Secondly, samples with alternate film stacks are analyzed. Specifically, for barriers deposited directly onto silicon, Cu diffusion is characterized through the formation of Cu3Si at elevated temperatures via detection by X-ray diffraction. Based on the formation temperatures of this Cu silicide, a comparison of the barrier stabilities is carried out. Furthermore, the barrier failure mechanisms are discussed and conclusions for a further improvement of the barrier performance are drawn.

Keywords: Diffusion barrier, Ta-TaN layer stacks, Ta-Si-N single layers, Cu interconnects, annealing, X-ray diffraction, glow-discharge optical emission spectroscopy, transmission electron microscopy, atomic absorption spectrometry, secondary ion mass spectroscopy

PACS: Replace this text with PACS numbers; choose from this list: http://www.aip.org/pacs/index.html.

Chapter 1

INTRODUCTION

1.1. Motivation

State-of-the-art high-performance microprocessors contain more than 100 million transistors on an area smaller than 200 mm^2. To connect these transistors and, thus, to be able to distribute clock and other signals between these active devices, a highly-complex wiring system is necessary. For on-chip interconnection, a three-dimensional network with an ever-growing number of metallization levels, which are divided by interlayer dielectrics, is applied. With continuous scaling down of the transistors, their influence on the delay in signal propagation is decreased compared to the impact of the metallization system. However, using sophisticated interconnect designs in combination with new technologies and materials, the so-called *RC* (resistance x capacitance) delay can be reduced, too. This task includes the implementation of interconnect materials with low resistivity and interlayer dielectrics with low permittivity. For this reason, Al-based interconnects have been already replaced by inlaid copper providing a higher electrical conductivity. Currently, much effort is undertaken to successfully implement insulator materials with a lower dielectric constant than silicon oxide.

Besides a lower resistivity (ρ_{Cu} = 1.673 µΩcm at T = 20 °C [1]) and, thus, a decreased delay time in signal propagation compared to Al interconnects [2], additional physical properties of copper are advantageous for microelectronic applications. Copper is characterized by a high thermal conductivity (k_{Cu} = 4.01 WK^{-1}cm^{-1} at T = 27 °C) [3]) preventing an excessive increase of the interconnect line temperature. The high melting point ($T_{m,Cu}$ = 1083 °C [1]) is responsible for a reduced influence of diffusion-controlled failure mechanisms at a particular temperature when compared to aluminum. Especially, a significantly improved electromigration behavior is observed for copper [4, 5]. However, the implementation of copper in microelectronic devices also causes severe problems. Since exposing to dry air does not lead to the formation of a protective oxide layer [6], Cu surfaces have to be passivated to minimize deleterious environmental influences. The diffusion of Cu ions in silicon and many dielectric materials under electrical and bias stress is particularly disadvantageous, as it causes drastic changes of the electrical properties of the microelectronic devices and finally leads to their degradation [7]. It should be mentioned that an increase of the Cu resistivity due to electron scattering from both grain boundaries and interfaces poses an additional critical challenge when the Cu interconnect cross-sections are further scaled-down [8, 9, 10].

To prevent Cu diffusion from interconnect lines into the interlayer dielectric as well as into the active transistor regions of microprocessors, effective diffusion barriers are indispensable. Regarding its application in microelectronic devices, the diffusion barrier B between the metallization layer M and the dielectric D has to satisfy the following requirements [11, 12, 13]:

- Defect-free microstructure of B and its stability under temperature stress,
- Low diffusion coefficients of M in B and of the components of D in B,
- High chemical stability between M and B as well as between D and B,
- High electrical conductivity of B,
- High thermal conductivity of B,
- Good adhesion of M to B and of B to D – particularly under temperature stress,
- Excellent mechanical properties of B (high stiffness and fracture toughness) – particularly in connection with low-k dielectrics,
- Ability of ultrathin and conformal deposition of B,
- Technological compatibility to the common back-end-of-line (BEOL) process.

Due to the continuous scaling down of the transistor feature sizes and, consequently, of the Cu interconnect dimensions, the thickness of the diffusion barriers has to be reduced, too. Thus, a significant increase of the interconnect line resistance can be prevented. According to the 2005 ITRS roadmap [14], the barrier layer thicknesses for Cu intermediate wiring in the 65 nm technology node will be 5.2 nm, whereas for the 14 nm node, the liner has to be not thicker than 1.1

nm. These requirements illustrate the efforts which have to be undertaken in the future to deposit ultrathin, conformal diffusion barriers with a defect-free and thermally stable microstructure onto mechanically weak low-k and ultra low-k interlayer dielectrics.

1.2. Diffusion in Thin Film Samples and Bulk Materials

1.2.1. Fick's First Law

Let $c_A(x,t)$ be the concentration distribution of A atoms in a binary AB alloy along the x-axis at given time t. Then, according to Fick's first law, there are two contributions to the mass flux $J_A(x,t)$ of A atoms [15]

$$J_A = -D_A \frac{\partial c_A}{\partial x} + c_A v_D \quad . \tag{1}$$

The first term in equation (1) corresponds to the mass flux caused by the Brownian migration of atoms. In this case, jumps of A atoms between two adjacent lattice planes are equiprobable in positive and in negative x-direction and the mass flux only arises due to different atomic occupation densities of adjacent lattice planes. Thus, the flux is proportional to the concentration gradient $\partial c_A/\partial x$, but directly opposite to it. The (Brownian) diffusion coefficient for component A is given by

$$D_A = \alpha \Gamma_S a^2 , \tag{2}$$

where Γ_S and a are the jump frequency and the jump distance of the atoms, respectively and α is a geometric factor [15]. The second term in equation (1) is caused by a driving force F and is called drift term. With F, the drift velocity v_D is given by

$$v_D = D_A \frac{F}{k_B T} \tag{3}$$

In the so-called Nernst-Einstein equation (3) k_B denotes the Boltzmann constant and T is the temperature. External driving forces F can be caused by a gradient of the electrostatic potential, a stress gradient, or a temperature gradient.

For a non-ideal solution, a gradient of the activity coefficient can give rise to a net drift term, too. In this case, jumps of atoms between adjacent lattice planes are not necessarily equiprobable in positive and in negative x-direction. With the activity coefficient γ_A of component A, which can be expressed by the quotient of the activity a_A and the mole numbers n_A, the chemical driving force F_{ch} is given by

$$F_{ch} = -k_B T \frac{\partial \ln \gamma_A}{\partial x}. \tag{4}$$

Thus, the mass flux J_A can be written as

$$J_A = -D_A \left[1 + \frac{\partial \ln \gamma_A}{\partial \ln n_A}\right] \frac{\partial c_A}{\partial x} = -D_A \Phi \frac{\partial c_A}{\partial x} = -D^A \frac{\partial c_A}{\partial x}, \tag{5}$$

where Φ is the thermodynamic factor and D^A is the intrinsic diffusion coefficient.

1.2.2. Bulk Diffusion – Fick's Second Law

In a non-steady one-dimensional case, the continuity equation for non-reacting particles can be written as

$$\frac{\partial c_A}{\partial t} + \frac{\partial J_A}{\partial x} = 0. \tag{6}$$

Inserting equation (5) into equation (6) gives Fick's second law

$$\frac{\partial c_A}{\partial t} = \frac{\partial}{\partial x}\left(D^A \frac{\partial c_A}{\partial x}\right). \tag{7}$$

For chemically homogeneous systems, the diffusion constant is independent of the concentration and equation (7) can be written as

$$\frac{\partial c_A}{\partial t} = D_A \frac{\partial^2 c_A}{\partial x^2}. \tag{8}$$

Using the boundary conditions $c_A(x,0) = M\delta(x)$ and $(\partial c_A/\partial x)(0,t) = 0$, the so-called *thin film solution* of equation (8) is given by

$$c_A(x,t) = \frac{M}{\sqrt{\pi D_A t}} \exp\left(-\frac{x^2}{4 D_A t}\right), \tag{9}$$

where M is the total amount of diffusant per unit area at time $t = 0$ [16]. Equation (9) is often used for the evaluation of diffusion experiments. Its application is, however, only valid when the thickness of the diffusant layer prepared onto a substrate is significantly smaller than the diffusion length $2(D_A t)^{1/2}$.

1.2.3. Grain Boundary Diffusion

To mathematically characterize the process of grain boundary diffusion, it is assumed that a grain boundary can be described by a homogeneous disk of thickness δ. The associated diffusion coefficient D' is significantly larger than the corresponding value D for bulk diffusion in both adjacent grains. Depending on the annealing time and the grain size d, Harrison [17] classified grain boundary diffusion into three kinetic regimes (Figure 1):

C-type kinetics: Diffusion in this regime is characterized by a bulk diffusion length $(Dt)^{1/2}$ which is significantly smaller than the grain boundary thickness δ. In this case, diffusion is restricted to the grain boundaries only. The determination of the diffusion coefficient D' can be carried out via recording of concentration profiles. However, for methods capable of evaluating the absolute concentration, the grain boundary volume fraction $g \approx \delta/d$ must be known.

A-type kinetics: Since the bulk diffusion length $(Dt)^{1/2}$ is significantly larger than the grain size d, there is an overlapping of the diffusion fields which developed around each boundary in the bulk. The diffusion process can be described according to Hart's equation with an effective diffusion coefficient D_{eff} [19]. For the determination of D_{eff}, the same methods can be applied as for the evaluation of the bulk diffusion coefficients.

B-type kinetics: This is the general case. Assuming a small thickness δ so that the concentration c' within the grain boundary is almost constant, a grain boundary diffusion coefficient D' which is significantly larger than the bulk diffusion coefficients D_1 and D_2 of both adjacent grains, as well as a continuous mass flux at the grain / grain boundary interface, the equations of the Fisher model [20] can be applied. For $D_1 = D_2$, Whipple [21] solved this diffusion problem based on a constant diffusant concentration at the surface, while Suzuoka

[22, 23] obtained the solution for an initially very thin surface layer of the diffusing species.

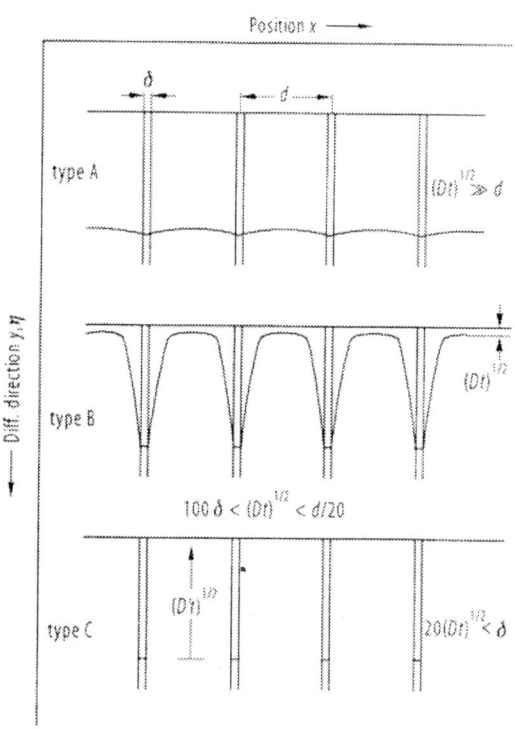

Reprinted from reference [18] with kind permission from Springer Science and Business Media.

Figure 1. Schematic representation of A, B, and C-type of grain boundary diffusion.

1.2.4. Dislocation Diffusion

Diffusion in dislocations can be characterized in an analogous way to grain boundary diffusion, whereby the dislocation is modeled as a homogenous cylinder with a particular radius [24]. Consequently, there are three diffusion regimes: In the case of C-type kinetics, the dislocation diffusion coefficient can be determined directly, while the Hart model has to be applied for A-type kinetics. In the case of C-type kinetics, the exact solution of the diffusion problem is as complex as for grain boundary diffusion. Defining $\overline{c}(y_0)$ as the laterally averaged concentration in the plane $y = y_0$ parallel to the surface, there is a linear relationship between $\ln \overline{c}$

and $y^{6/5}$ in the case of grain boundary diffusion [21, 22, 23], whereas $\ln \bar{c}$ varies linearly with y for dislocation diffusion [24].

1.2.5. Temperature dependence of diffusion

Considering the atomic diffusion mechanisms, e.g. the vacancy mechanism, the jump frequency Γ_S introduced in equation (2) is given by the product of the defect fraction n, the exchange frequency ω between atom and defect, and a correlation factor f. Thus, the temperature dependence of the diffusion coefficient is determined by the latter three parameters. The defect creation as well as the exchange between atom and defect can be treated as thermally activated processes. Consequently, with $q = n, \omega$, it is

$$q \propto \exp\left(\frac{\Delta S_q}{k_B}\right) \exp\left(-\frac{\Delta H_q}{k_B T}\right), \tag{10}$$

where ΔS_q and ΔH_q are the activation entropy and the activation enthalpy of the corresponding process, respectively. In the case of a temperature-independent correlation factor f, the diffusion coefficient D is consequently characterized by the following exponential temperature dependence

$$D = D_0 \exp\left(-\frac{Q}{k_B T}\right), \tag{11}$$

where D_0 is the pre-exponential factor and Q is the activation energy. For the vacancy mechanism, Q is given by the sum of the enthalpies for the formation and the migration of the vacancies.

1.2.6. Diffusion hierarchy

Depending on the thermal conditions, there are principally three different kinds of defects which contribute to the diffusion in thin metal films and bulk samples, namely point defects, dislocations, and grain boundaries. Dislocations and grain boundaries themselves contain vacancies and interstitials. Consequently, atom-vacancy exchanges and the direct migration of interstitial atoms are the most important processes contributing to diffusion. The formation and motion of point defects are, however, different in the crystal lattice and along dislocations and grain boundaries. This fact results in a hierarchy of diffusion rates first proposed by Gjostein [25]. In particular, the activation energies for the various diffusion processes differ being largest for bulk diffusion and smallest for surface diffusion.

According to Gjostein [25], there are direct linear relationships between the activation energy for a particular diffusion process and the melting temperature of the metallic solid. For bulk diffusion, the activation energy is given by Q [cal/mol] = 34 T_m [K], while it can be written as Q' [cal/mol] = 20 T_m [K] for grain boundary diffusion.

1.3. Diffusion of Copper in Silicon and Dielectric Materials

The fast diffusion of copper in silicon and in most dielectric materials has turned out as one of the most serious problems for the application of copper as interconnect material in microprocessors. In Si single crystals, Cu atoms diffuse as interstitials. In the temperature range 0 °C < T < 900 °C, this diffusion process is characterized by pre-exponential factor $D_{0,Cu/Si} = 3.0*10^{-4}$ cm^{2}s^{-1} and activation energy $Q_{Cu/Si} = 0.18$ eV [26]. Cu atoms occupying interstitial lattice sites in silicon act as shallow single donors [27], while substitutional copper is a triple acceptor [27, 28]. The ratio of substitutional to interstitial copper is given by 10^{-4} [27, 29]. Due to Coulomb interaction between differently charged ions, the formation of Cu-acceptor pairs results in a covalent bonding of copper to a dopant, such as Al, In, Ga, or B [30]. In the case of boron, Cu-acceptor pairing can lead to a complete acceptor passivation [31]. The solubility of copper in silicon is $S_{Cu/Si} = 10^{18}$ cm^{-3} at a temperature of $T = 1000$ °C, while the extrapolation to $T = 25$ °C leads to $S_{Cu/Si} < 1$ cm^{-3} [29]. Consequently, cooling down a Si single crystal containing copper results in the formation of defects which can be classified into five groups: (*i*) formation of point defects and their complexes in the bulk, (*ii*) formation of Cu silicide precipitates in the bulk, nucleated either homogeneously or, more likely, at microscopic lattice defects, (*iii*) decoration of existing extended defects such as dislocations, grain boundaries, etc., (*iv*) outdiffusion to the surface, and (*v*) segregation in p$^+$ areas [7]. In the case of a Cu layer deposited directly onto a Si wafer, the formation of Cu$_3$Si starts at $T \approx 190$ °C which can be deferred from Figure 2. In particular, the two-dimensional representation $I(2\theta, T)$ of X-ray patterns recorded during heating up a Cu (200 nm) / Si sample shows that the intensities for both observed Cu diffraction maximums start to decrease at $T \approx 190$ °C. At the same temperature, additional Bragg reflections, which can be assigned to Cu$_3$Si, appear. Besides the low reaction temperature [32, 33], the fact that the Cu silicide formation is easily detected via X-ray diffraction can be used for a sensitive proof of Cu diffusion through a diffusion barrier. It should be mentioned that copper is the diffusant in the transport-controlled reaction with silicon [34, 35]. Although the crystal structure of Cu$_3$Si has not been unambiguously clarified yet [36, 37], detailed studies about its growth behavior exist [38, 39, 40]. Even at room temperature, the presence of Cu_3Si catalyzes the oxidation of silicon [41, 42, 43].

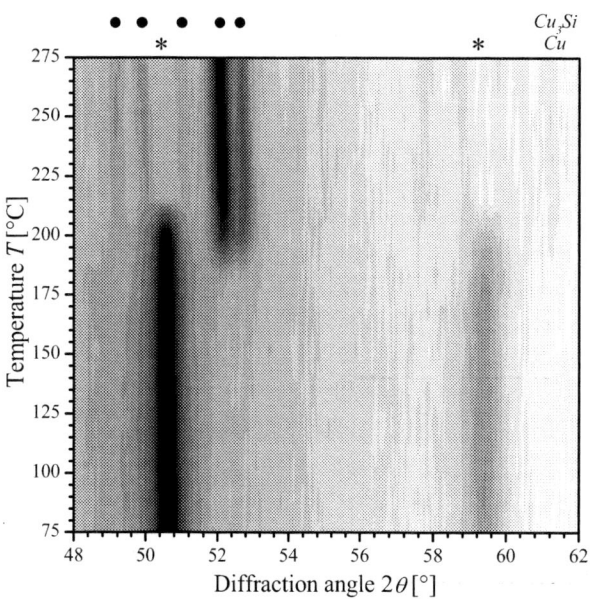

Figure 2. Two-dimensional representation $I(2\theta, T)$ of X-ray patterns recorded during heating up a Cu (200 nm) / Si sample with a rate of 4 Kmin^{-1}. The in situ X-ray experiment was carried out using a position-sensitive detector and Co-K$_\alpha$ radiation at fixed incidence angle.

Copper diffusion is not only fast in silicon, but also in most dielectric materials relevant to microelectronics. For a Cu layer deposited directly onto thermal oxide and annealed in argon, significant thermal diffusion is observed at $T = 400$ °C [44], while it occurs at $T > 400$ °C if the heat treatment is performed in Ar-3 v.% H_2 [45]. Thus, a small amount of oxygen in the annealing ambience seems to accelerate Cu diffusion [46]. Compared to thermal oxide, Cu diffusion is slower in silicon nitride [47]. However, the Cu solubility is higher in the latter material [47]. Under electrical bias, the drift of Cu ions in dielectrics is significantly enlarged [48]. According to Pai et al. [49] and Vogt et al. [50], silicon nitride layers show a higher thermal stability than silicon oxide layers. Applying an electrical field with a strength of $E = 4$ MVcm^{-1}, Raghavan et al. [51] determined an activation energy of $Q_{Cu/SiO_2} = 1.2$ eV for the Cu ion drift in thermal oxide. At $E = 1$ MVcm^{-1} and 150 °C $< T <$ 300 °C, Shacham-Diamand [52] characterized the same process by $D_{0, Cu/SiO_2} = 2.5*10^{-8}$ cm^{2}s^{-1} and $Q_{Cu/SiO_2} = 0.93$ eV. According to Wendt et al. [53], the low Cu solubility in silicon and SiO$_2$ coupled with the high mobility in both materials can lead to segregation of Cu-rich silicides at the Si/SiO$_2$ interface.

1.4. Diffusion Barriers for Cu Interconnects

1.4.1. Potential Materials for Cu Diffusion Barriers

To prevent diffusion of Cu atoms from the on-chip interconnect lines into the adjacent interlayer dielectric, diffusion barriers are needed. Besides a high stability against Cu diffusion, the barriers have to meet further electrical, thermal, chemical, and mechanical requirements, as described above. Considering the continuous scaling down of the interconnect cross-sections and the implementation of low-k and ultra low-k dielectrics, ultrathin diffusion barriers deposited with high conformality have to contribute to a sufficient mechanical stability of the Cu / barrier / low-k dielectric system and have to provide excellent interface adhesion. Besides the decision for a specific barrier material, the choice of the deposition method is crucial. It has a large influence on the barrier microstructure and, thus, on the type and density of the intrinsic barrier defects which determine the diffusion behavior.

Single crystalline layers of a suitable material should act as the best diffusion barriers. Due to the complex structure of the three-dimensional metallization system with structurally and chemically varying interfaces, a single crystalline and, consequently, defect-free film growth is, however, difficult to realize, even with new deposition techniques, such as atomic layer deposition (ALD). Polycrystalline, nanocrystalline, and amorphous diffusion barriers can be deposited much easier than single crystalline layers. The microstructure of polycrystalline films depends significantly on the choice of the material and the deposition conditions [54, 55, 56]. Columnar grain growth, which is often observed for polycrystalline layers, leads, however, to a high density of grain boundaries and, consequently, to the formation of fast diffusion paths which is disadvantageous for an application as a diffusion barrier. In the case of nanocrystalline films, the grain size is usually much smaller than the layer thickness which results in an elongation of the diffusion paths. This advantage might, however, be lost, if the barrier thickness continues to decrease. Then, the application of amorphous films would be the only possibility. However, the amorphous microstructure must not change during different kinds of stressing. In particular, barrier crystallization and, consequently, grain boundary formation should not occur at annealing.

According to the empirical rule that the activation energy for a specific diffusion process is proportional to the material's melting temperature, diffusion barriers should be characterized by a high melting point T_m. For a particular temperature, the influence of a specific diffusion mechanism (e.g. grain boundary diffusion) is thus reduced compared to materials with low T_m. Consequently, refractory metals with melting points that are higher than those of iron, cobalt, and nickel [57] as well as binary and ternary compounds containing refractory metals are suitable candidates as diffusion barriers for Cu interconnects. They can be found among the following five groups of materials which are currently investigated for this purpose:

(1) Polycrystalline refractory metals (Me), such as Cr [58], Ti [59], Mo [60], W [61], Ta [62], Nb [63],
(2) Polycrystalline and amorphous Me alloys, such as TiW_x [64], $TaCo_x$ [65], $TaFe_x$ [65], TaW_x [66], $NiNb_x$ [67], $CuZr_x$ [68],
(3) Polycrystalline and amorphous Me-Si compounds, such as $MoSi_x$ [69], WSi_x [70], $TaSi_x$ [71],
(4) Polycrystalline and amorphous Me-N, Me-C, Me-O, and Me-B compounds, such as TiN_x [72], VN_x [73], ZrN_x [74], NbN_x [75], MoN_x [60], HfN_x [76], WN_x [77], TaN_x [78], WC_x [79], TaC_x [80], MoO_x [81], TaO_x [82], TiB_2 [83],
(5) Amorphous ternary Me compounds, such as $TiSi_xN_y$ [84], $MoSi_xN_y$ [69], WSi_xN_y [85], $TaSi_xN_y$ [86], TaW_xN_y [87], CoW_xP_y [88], $TiAl_xN_y$ [89], WB_xN_y [90], WC_xN_y [91], WGe_xN_y [92], ZrC_xN_y [93].

Although ternary compounds already offer many possibilities to change the barrier properties, recent studies suggest the application of quaternary compounds, such as $TiAl_xN_yO_z$ [94] as well as $CoW_xRe_yP_z$ and $CoW_xRe_yB_z$ [95]. Organic-based diffusion barriers are another interesting approach discussed in literature. According to Senkevich et al. [96], bias-temperature stressing (BTS) of a metal-insulator-semiconductor capacitor with a 5 nm thick polymer (parylene-N) diffusion barrier at $T = 150$ °C, $E = 1$ MV/cm and previously annealed at $T = 250$ °C in Ar-3%H_2 did not result in a flatband voltage shift which indicates a stable polymer-capped dielectric. Ganesan et al. [97] demonstrated the use of carboxyl- and amine-terminated self-assembled molecular layers (SAMs) to immobilize Cu at the Cu/SiO_2 interface. Amine-terminated SAMs at the Cu/SiO_2 interface increase the Cu diffusion-induced device failure time by a factor of 3 compared to interfaces without a barrier. Carboxyl-terminated SAMs show an increase in failure time by more than an additional factor of 4 [97].

Nowadays, void-free Cu filling of trenches and vias is done by electrochemical deposition (ECD). For this process, a continuous and smooth film is required to carry the current [98]. As most of the above-mentioned diffusion barrier materials are characterized by a resistivity which is too high to directly deposit ECD-Cu, an ultrathin nucleation or seed layer has to be sputtered on top of the barrier. Copper prepared by physical vapor deposition (PVD) can be used for this purpose. With continuous scaling down of all feature sizes, the interconnect cross-sections must shrink, too. Consequently, it will become difficult to deposit the diffusion barrier and the Cu seed layer with conventional PVD techniques. That is why it would be desirable to have a Cu-plateable diffusion barrier. Lane et al. [99] identified a family of metals which allow for direct Cu electrodeposition in standard acid plating baths. Ruthenium with a bulk resistivity of ρ_{Ru} = 7.1 µΩ cm is one of the metals proposed. It is stable in air and shows excellent Cu wettability [100] as well as negligible Cu solubility, even at T = 900 °C [101]. Chyan et al. [102] and Chan et al. [103] demonstrated that ruthenium is an excellent substrate for conformal Cu electroplating. The inherently favorable interface bonding manifests in the strong adhesion between ECD-Cu and ruthenium even after annealing at T = 600 °C [102]. Josell et al. [104] reported seedless superfilling of Ru-coated trenches by direct Cu plating. Concerning the thermal stability, Arunagiri et al. [105, 106] showed 5 nm Ru films deposited between copper and silicon to be reliable barriers up to at least T = 300 °C vacuum anneal. Heat treatment at T = 450 °C leads, however, to barrier failure which seems to be caused by Ru silicide formation at the Ru/Si interface [105, 106]. According to Damayanti et al. [107], N addition during Ru barrier sputter deposition results in an improvement of the thermal stability. Detectable Cu diffusion into SiO_2 was not observed until T = 500 °C. It should, however, be mentioned that the predominantly amorphous 10 nm thick Ru-N barrier already releases most of its nitrogen during annealing at T = 275 °C which consequently leads to the formation of hexagonal ruthenium [107]. Besides Ru-based films, Ir layers are investigated as Cu-plateable diffusion barriers. In particular, Josell et al. [108] demonstrated seedless superfilling by Cu electrodeposition on sub-micrometer trenches with Ir barrier layers.

As an alternative to the conventional barrier process, the implementation of so-called self-forming diffusion barriers has been proposed. In this case, a Cu alloy film is directly deposited onto the interlayer dielectric and subsequently annealed so that the alloying element can migrate to the Cu alloy / dielectric interface. There, a barrier layer is formed via a reaction with the dielectric. Tsukimoto et al. [109] investigated a self-forming barrier layer based on Cu(Ti), while Chu et al. [110] suggested a co-sputtering of copper and tungsten in an Ar/N_2 gas mixture followed by a heat treatment at T = 530 °C for t = 1 h. According to Koike et al. [111], an only 3-4 nm thick Mn-containing diffusion barrier forms at the Cu(Mn)/SiO_2 interface during annealing at T = 450 °C for t = 30 min.

1.4.2. Barrier Deposition Techniques

So far, Cu diffusion barriers and seed layers have been deposited by PVD. Due to intensive technological improvements for an enhancement of the film conformality, this deposition method is applicable at least up to the 65 nm technology node by using ionized metal plasma (IMP) [112] and self-ionized plasma (SIP) [113] techniques. In the case of IMP, the metal ions are generated by radio frequency power. Ionized PVD (IPVD) films exhibit the low resistivity, high density, and good adhesion to the underlying dielectric desired for an application as a diffusion barrier. For the SIP approach, the magnetron source is designed to provide an efficient energy transfer from the secondary electrons to the plasma waves which leads to increased ionization of the sputtered metal atoms. In both cases, the flux of the metal ions and the amount of re-sputtering can be controlled by the wafer bias. Overhang at the top corner of a trench or via structure, step coverage, and bottom re-sputtering are determined by the directionality of ionized metal atoms which again is controlled by the wafer bias [13].

In the near future, especially for the 45 nm technology and beyond, very conformal, ultrathin film deposition technique will be required. Layers deposited by IPVD may not provide sufficient step coverage to reliably coat features having a high aspect ratio and sub-100 nm dimensions. Alternatively, chemical vapor deposition (CVD) can be used to prepare highly conformal metal films, but the electrical performance and interface quality may not equal that of PVD layers. To address future diffusion barrier deposition needs, Li et al. [114] reported a method providing PVD-like film quality and CVD-like step coverage. In this so-called chemically enhanced physical vapor deposition (CEPVD), a chemical precursor is introduced at the substrate during IPVD to provide a CVD component to the overall deposition process. In particular, Li et al. [114] deposited Ta-N based films by reactive sputtering of a Ta target in Ar/N_2 plasma and the presence of a Ta-containing organometallic precursor. The obtained films are characterized by a low resistivity and improved step coverage compared to IPVD layers, but contain an appreciable amount of carbon (30–60 at.% depending on process conditions) [114].

Due to the ability to deposit highly conformal films onto small features, atomic layer deposition is considered as one of the most promising deposition techniques in future technology nodes. The process is intrinsically atomic in nature and results in the controlled deposition of films in sub-monolayer units. Ideally, film thickness is determined by the number of deposition cycles, rather than the timing of a continuous deposition process (like PVD) with a precalibrated deposition rate [115]. In ALD, one deposition cycle consists of the following steps: After exposing the substrate to a metal precursor vapor carried by a carrier gas (usually Ar) for a given time, the chamber is evacuated. Then, the reactant gases are introduced for a set time. In the case of plasma-enhanced ALD (PEALD), a RF plasma is additionally initiated. In the last step, the chamber is pumped again and the base pressure is reestablished. Different diffusion barrier materials, such as MoN_x [116], NbN_x [75], TiN_x [117], WN_x [118], and WC_xN_y [91] have been prepared by ALD. In particular, interconnect structures fabricated using ALD-WC_xN_y barriers and Aurora low-k dielectric layers in a standard BEOL process flow demonstrated electrical properties superior to a reference PVD-Ta/TaN bilayer [91]. Ta-based films can be also deposited by ALD. For this purpose, there are principally two types of metal precursors: (i) Ta halides, such as $TaCl_5$, $TaBr_5$, and TaI_5 and (ii) metal organic precursors, such as pentakis(dimethylamino)tantalum (PDMAT), pentakis(diethylamino)tantalum (PDEAT), tert-butylimidotris(diethylamido)tantalum (TBTDET), and tert-amylimidotris(dimethylamido)tantalum (TAIMATA). The deposition of ALD-Ta films can be only performed using Ta halides and a reducing agent, such as hydrogen [119], because all currently available metal organic precursors contain nitrogen. On the contrary, the preparation of Ta-N based layers using thermal ALD or PEALD is reported for both precursors Ta halides [120, 121] and metal organic compounds [122, 123]. Ru thin films can be also deposited by ALD. For this purpose, the metal organic precursors bis(cyclopentadienyl)ruthenium and $Ru(Od)_3$ [Od=2,4-octanedionate] were used in an oxygen ambience by Aaltonen et al. [124] and Min et al. [125], respectively. Kwon et al. [126] demonstrated plasma-enhanced ALD of polycrystalline Ru layers by a reaction of bis(ethylcyclopentadienyl)ruthenium [$Ru(EtCp)_2$] and NH_3. However, to promote a uniform nucleation of ruthenium on an oxide surface, the deposition of an additional layer seems to be necessary [124, 126]. Using a thin PEALD-TaN nucleation film, Kim [115] reported conformal deposition of ALD-Ru into trench structures without TaN penetration into the porous low-k material. However, since the Ru barrier failure occurred at about 700 °C, which is similar to a 5 Å thick

TaN layer, it was concluded that the Ru film itself is not an effective diffusion barrier [115].

1.4.3. Ta-Based Diffusion Barriers

According to research studies done so far, the refractory metal tantalum as well as binary and ternary Ta-based materials have been shown to be particularly effective diffusion barriers for Cu metallization. By presenting selected results concerning microstructure, physical properties, thermal stability, and failure mechanisms of such diffusion barriers, this will be demonstrated below.

Ta Diffusion Barriers

The refractory metal tantalum (melting temperature: T_m = 3020 °C [127]) satisfies important requirements for an application as diffusion barrier material. Intermetallic Cu-Ta compounds do not exist [127, 128] and the solid solubility of tantalum in copper and vice versa is very small [127]. According to angular resolved X-ray photoelectron spectroscopy (ARXPS) investigations, the Ta/Si and the Ta/SiO$_2$ interfaces have a layered internal structure [129]. Both interfaces are thermally very stable. Tantalum does not react with silicon until T = 625 °C [130] and a significant reaction with silicon oxide only occurs for $T \geq$ 1100 °C [131]. Tantalum shows strong adhesion to copper [132] which can be further improved by inserting an ultrathin layer of poly(acrylic acid) [133]. However, the adhesion of tantalum to silicon oxide is worse when compared with the Ta nitride adhesion to the same dielectric [134]. So far, three different Ta modifications have been mainly observed: (i) thermodynamically stable body centered cubic (bcc) α-Ta (space group: $I\bar{m}3m$ (229), lattice parameter (single crystal): a = 5.43010(5) Å [135, 136], resistivity (layer): $\rho_{\alpha\text{-}Ta}$ = 15 ... 50 μΩcm [137]), (ii) thermodynamically metastable tetragonal β-Ta (space group: $P\bar{4}2_1m$ (113), lattice parameters (single crystal): a = 10.211(3) Å, c = 5.3064(10) Å [138], resistivity (layer): $\rho_{\beta\text{-}Ta}$ = 150 ... 220 μΩcm [137]), and (iii) amorphous Ta (resistivity (layer): $\rho_{a\text{-}Ta}$ = 150 ... 180 μΩcm [119]). Depending on the deposition method and the deposition parameters, a Ta layer contains one or a mixture of these phases. At room temperature, metastable β-Ta usually grows on silicon and SiO$_2$ [130, 139, 140], while α-Ta is deposited at elevated substrate temperatures (T > 400 °C) [141]. The thermodynamically stable modification will be also observed if tantalum is prepared onto thin (>0.3 nm) Nb [142] or Ti films [143]. Annealing experiments can be used to transform metastable β-Ta into stable α-Ta, whereby the transformation temperature is significantly determined by the deposition conditions. For a 100 nm thick Ta layer with intrinsic compressive stress, the transformation occurs abruptly at $T = 750$ °C and is accompanied by a stress relief [144, 145]. According to Hoogeveen et al. [146], the transformation temperature decreases with increasing Ta layer thickness. Ultrathin amorphous Ta films can be prepared by atomic layer deposition [119].

Due to the above-mentioned properties, thin Ta films have been intensively investigated as diffusion barriers between copper and silicon in the last few years [147, 148]. Polycrystalline Ta layers are characterized by a high thermal stability which seems to be independent of the Ta crystal structure [148, 149]. Holloway et al. [140] detected significant Cu diffusion through a 50 nm thick Ta barrier together with $TaSi_2$ formation after a $T = 600$ °C / $t = ½$ h anneal. In a later study [150], they concluded that Cu diffusion along Ta grain boundaries into the Si substrate promotes and, thus, precedes the nucleation of Ta silicide. According to Laurila et al. [62], the order in which Cu and Ta silicide formation occurs depends on the thickness of the diffusion barrier. Thick Ta layers (~ 100 nm) are able to prevent Cu diffusion into silicon even at the $TaSi_2$ formation temperature. In the case of thin Ta films (10 ... 50 nm), however, Cu atoms can penetrate the barrier before $TaSi_2$ nucleation and react with silicon to form Cu_3Si [62]. The dependence of the thermal stability of 20 nm thick polycrystalline Ta diffusion barriers on the residual gas pressure during deposition was investigated by Clevenger et al. [151]. According to this study, Ta layers deposited under UHV conditions showed a broad range of failure temperature (320 °C $< T <$ 630 °C), while those Ta films prepared under HV conditions failed between $T = 600$ °C und 630 °C. The narrow temperature range in the latter case was explained with unintentional incorporation of oxygen during barrier deposition. Furthermore, Clevenger et al. [151] showed that below an O concentration of 0.5 ... 1.0 at.%, the barrier failure is mainly determined by the residual H concentration as well as microstructure defects which allow for fast Cu diffusion through tantalum. The Ta barrier stability depends not only on the vacuum base pressure during deposition, but also on the residual gas pressure during annealing. Yin et al. [152] demonstrated that heat treatment in vacuum at pressure $p_0 \approx 1$ Pa leads to O diffusion along Cu grain boundaries into the Cu/Ta interface and, consequently, to the formation of a thin amorphous Ta oxide layer. According to Laurila et al. [153], such an interface oxide layer acts as an additional barrier against Cu diffusion. Various attempts have been undertaken to further improve the thermal stability of Ta diffusion barriers. Kang et al. [154] deposited Ta layers under Ar ion bombardment which results in a higher packing density of the grain boundaries and, thus, in an increased barrier stability. Further possibilities include the incorporation of CeO_2 into Ta grain boundaries [155] as well as the insertion of an additional Zr [156] or

V film [157] into the Ta barrier. Using ALD, the deposition of amorphous Ta films is possible [119]. In comparison to PVD-Ta films of similar thickness, ALD-Ta layers are characterized by a failure temperature which is at least 70 K higher [119].

Ta diffusion barriers show a significantly higher thermal stability when deposited onto SiO_2 [158, 159, 160]. According to Wang et al. [160], there is no Ta compound formation for a 25 nm thick Ta layer between copper and SiO_2 even after annealing at $T = 800$ °C. However, for the same Ta barrier prepared onto silicon, Ta silicide formation occurs at $T = 650$ °C [160]. Although Jang et al. [161] also did not observe any phase formation in the Cu/Ta/SiO_2/Si system after heat treatment at $T = 800$ °C for $t = ½$ h, they noticed a diffusion of Ta atoms through the Cu layer to the sample surface. A further investigation of the consequences of such Ta diffusion on the morphology and the structure of the remaining barrier layer was, however, not performed. Due to its relatively high permittivity, silicon oxide will continue to be replaced by low-k and ultra low-k interlayer dielectrics. Thus, a further decrease of the metallization-induced RC delay in signal propagation can be achieved. Currently, dense low-k materials, such as organosilicate glass or spin on dielectrics with $k \approx 2.5$ are used. However, for a further reduction of the effective dielectric constant, the implementation of porous ultra low-k dielectrics with $k < 2.0$ seems to be necessary. This poses many problems, as porous dielectrics are characterized by inherent weak mechanical properties, a high leakage current, and a low thermal conductivity. In addition, thin Ta-based barriers deposited with conventional PVD do not satisfactorily seal the porous low-k sidewall surfaces [162]. To achieve better barrier performance and better electrical characteristics of Cu / porous ultra low-k interconnect structures, the deposition of a thin amorphous hydrogenated SiC_x layer between a PVD-Ta barrier and the porous low-k interlayer dielectric was proposed [163]. Ta/dielectric bilayer diffusion barriers with different silicon carbide films (SiC_x, SiC_xN_y, SiC_xO_y) were studied by Chen et al. [164]. These barriers show significant performance improvements in terms of breakdown strength and leakage current characteristics compared to conventional PVD-Ta barriers. According to the study, Ta/SiC_x and Ta/SiC_xN_y bilayers are potential candidates for an application in Cu / porous low-k damascene interconnects due to their superior electrical performance even after long-term BTS experiments [164].

Ta-N Diffusion Barriers

Due to the polycrystalline structure of α-Ta and β-Ta, Cu diffusion along grain boundaries is observed at elevated temperatures. Based on the pre-exponential factor $D^*_{0,Cu/Ta} = 9.0*10^{-4}$ cm^2s^{-1} and the activation energy

$Q'_{Cu/Ta}$ = 2.3 eV [165] for this diffusion process, the characteristic length $l = (Dt)^{1/2}$ is calculated to $l_{Cu/Ta} \approx 4$ nm for a given temperature of $T = 600$ °C and time $t = 1$ h. Besides the possibilities discussed above, the barrier stability can be also increased by N incorporation into the Ta films. To study the phase composition and microstructure of such Ta-N layers depending on the N content, detailed investigations have been carried out [166, 167, 168, 169]. According to Stavrev et al. [167], small additions of nitrogen to the sputter gas argon lead to a transformation of metastable tetragonal β-Ta into nanocrystalline bcc Ta(N), while further increase of the N$_2$ flow results in the formation of nanocrystalline face centered cubic (fcc) TaN. Stoichiometric fcc TaN (space group: $F\bar{m}3m$ (225), lattice parameter (powder): $a = 4.3363(1)$ Å [170]) is characterized by a strong adhesion to SiO$_2$ [134] and a high melting temperature ($T_m = 3090$ °C [171]). However, the resistivity of fcc TaN layers is slightly higher compared to Ta films [167]. The deposition of fcc ALD-TaN has been rarely successful. Using TaCl$_5$ as precursor and NH$_3$, high-resistivity Ta$_3$N$_5$ films were obtained and the use of hydrazine as reducing agent produced similar results [172]. Low-resistivity fcc TaN films with $\rho_{TaN} \approx 1500$ μΩcm but high impurity incorporation were prepared using additional reducing agents like Zn, trimethylaluminum, or amines [173, 174, 175]. However, Kim et al. [121] were successful in depositing low-resistivity ($\rho_{TaN} \approx 350$ μΩcm) fcc TaN films by PEALD using TaCl$_5$ and a plasma of both nitrogen and hydrogen.

Investigations concerning the influence of the N content on the diffusion barrier properties of Ta-N films between copper and silicon show that the thermal stability increases with increasing x_N/x_{Ta} ratio [78, 160, 176, 177]. Based on X-ray diffraction experiments, Wang et al. [160] detected the formation of Ta silicide and Cu silicide for 25 nm thick Ta and TaN barriers after annealing at $T = 650$ °C and 800 °C, respectively. According to atomic absorption spectrometry (AAS) studies done by Hecker et al. [78], Cu trace diffusion through 10 nm thick Ta, Ta$_{20}$N$_{80}$, and TaN barriers was observed for the first time after one hour heat treatment at $T = 450$ °C, 550 °C, and 600 °C, respectively. Kim et al. [178] showed that TaN films prepared by PEALD have diffusion barrier properties which are comparable to those of PVD-TaN layers. Investigations regarding the failure mechanisms of amorphous Ta-N barriers with Ta$_2$N compositions were conducted by Min et al. [179] and Chen et al. [180]. Caused by a crystallization process commencing at $T = 450$ °C as well as thermally induced microcracks, Cu atoms diffuse into silicon and react with it to form Cu$_3$Si. A slightly higher N content in the barrier leads to an increase of the crystallization temperature and a retardation of the subsequent formation of Cu$_3$Si and TaSi$_2$ [180]. Regarding TaN, the barrier failure is determined by Cu diffusion into the Si substrate [179]. In this case, the diffusion process is controlled by both the TaN grain boundaries and the TaN lattice with associated activation energies $Q'_{Cu/TaN} = 1.3$ eV and $Q_{Cu/TaN} = 2.7$ eV [181]. For the temperature range 500 °C $< T <$ 800 °C, Lin et al. [182] provided the following diffusion parameters: $D'_{0,Cu/TaN} = 2.36*10^{-11}$ cm^{2}s^{-1} and $Q'_{Cu/TaN} = 0.80$ eV for grain boundary diffusion as well as $D_{0,Cu/TaN} = 2.89*10^{-17}$ cm^{2}s^{-1} and $Q_{Cu/TaN} = 0.12$ eV for bulk diffusion.

Multilayered Ta/Ta-N Diffusion Barriers

Based on the research results obtained for Ta and Ta-N films, a combination of these single layers promises further optimization of the diffusion barrier properties [183, 184]. Chen et al. [184] investigated multilayers consisting of amorphous Ta$_2$N and nanocrystalline TaN films with a period thickness of 10 nm. They showed that even after Ta$_2$N crystallization, these multilayers better prevent Cu diffusion than each single layer of comparable thickness (40 nm).

There are at least two advantages for an application of a bilayer as a diffusion barrier. Firstly, the two different barrier sides allow for an optimization of physical properties at both adjacent interfaces. Secondly, the bilayer should be characterized by a thermal stability which is at least as high as for the more stable single layer. For a successful application of a bilayer diffusion barrier, there is, however, the precondition of a chemically stable interface between both single layers. The implementation of a Ta/TaN layer stack as Cu diffusion barrier was proposed in [185, 186]. Choosing appropriate deposition parameters, Edelstein et al. [185] were able to deposit hexagonal close-packed (hcp) TaN with only a small fraction of the cubic modification. Compared to fcc TaN, films of hcp Ta nitride are characterized by a lower intrinsic stress, a lower resistivity, and a comparably strong adhesion to SiO$_2$. The Ta layer, which adheres well to copper, consists of the electrically more conducting α-Ta modification, which forms during deposition onto TaN [185]. Using blanket substrates, Traving et al. [187] demonstrated that the TaN layer can be as thin as 0.7 nm to grow the α-Ta phase onto Ta nitride. In the case of trench structures, however, the nominal TaN thickness has to be increased with decreasing feature sizes to obtain pure α-Ta [187]. According to Ho et al. [188], a Ta/TaN bilayer is a suitable diffusion barrier down to the 65 nm technology node. In particular, they reported the successful integration of a PVD-Ta/ALD-TaN bilayer into a dual-damascene Cu/low-k interconnect structure with via sizes of 110 nm and an achieved RC reduction of up to 16 % relative to the PVD baseline [188].

Ta-Si-N Diffusion Barriers

Although Stavrev et al. [167] demonstrated that a higher N content in a Ta-N layer results in the formation of smaller Ta nitride grains, an amorphous diffusion barrier cannot be produced based on this approach. This is, however, possible by incorporating silicon into tantalum [189, 190, 191]. According to Lee et al. [71], the failure of such amorphous Ta-Si layers deposited between copper and silicon is determined by two mechanisms: (*i*) barrier crystallization into Ta silicides and (*ii*) Cu_3Si formation caused by a reaction of copper with silicon from the barrier. Kolawa et al. [192] showed that the thermal stability of Ta-Si layers is higher compared to pure Ta films. Si n^+p shallow junction diodes with Cu metallization and an integrated 180 nm thick Ta layer failed during annealing at $T = 550$ °C for $t = ½$ h, while those diodes with a 100 nm thick $Ta_{74}Si_{26}$ diffusion barrier remained intact even during heat treatment at $T = 600$ °C for $t = ½$ h [192].

With regard to binary Ta-N diffusion barriers, a further improvement of the thermal stability of Ta-Si layers seems to be possible by additional incorporation of nitrogen into the barrier [193, 194, 195]. Besides a high resistance against oxidation [196, 197] and a low intrinsic film stress [198, 199], PVD-Ta-Si-N barriers show a better adhesion to copper, promote a stronger <111> Cu texture component, and result in an improved electromigration behavior compared to Ta and TaN films [200, 201]. Furthermore, Ta-Si-N barriers are characterized by a high thermal stability between copper and silicon [202, 203, 204, 205] as well as between copper and SiO_2 [206]. According to Kolawa et al. [192], shallow junction diodes with a Cu (500 nm) / $Ta_{36}Si_{14}N_{50}$ (120 nm) / Si metallization are thermally stable up to annealing at $T = 900$ °C for $t = ½$ h. Barrier failure at higher temperatures was explained by accelerated $Ta_{36}Si_{14}N_{50}$ layer crystallization due to the presence of copper and subsequent fast diffusion of Cu atoms along grain boundaries into silicon. Until now, there are, however, only few systematic studies regarding the dependence of the thermal stability and the failure mechanisms of ternary Ta-Si-N diffusion barriers on their chemical composition [207, 208]. This is particularly true for barriers deposited between copper and an interlayer dielectric.

1.5. Aim of this Study

The preceding literature review clearly demonstrated that due to the fast Cu diffusion in silicon and most dielectric materials and the resulting deleterious changes in their electronic structure, the implementation of Cu diffusion barriers is indispensable. Furthermore, it was shown that refractory metals and their compounds are suitable barrier materials. In particular, Ta/TaN layer stacks and Ta-Si-N single layers show excellent properties making these diffusion barriers high-potential candidates for an implementation in future Cu interconnect structures.

Previous investigations of Ta-based diffusion barriers have mainly concentrated on the thermal stability and its dependence on the barrier composition without placing much emphasis on a detailed analysis of the corresponding failure mechanism. Most studies have been carried out for barrier layers deposited with thicknesses larger than 10 nm onto Si substrates. However, for an implementation in microelectronic devices, diffusion barriers prepared between copper and an interlayer dielectric are more relevant. Regarding the implementation of low-k and ultra low-k dielectrics as well as the continuous scaling down of all interconnect feature sizes, the microstructure of the diffusion barriers becomes more and more important. Defects, like grain boundaries or pinholes can act as fast paths for Cu diffusion and consequently lead to barrier failure.

It is the aim of this study to use complementary analytical methods for microstructure and functional property analyses of sophisticated Ta-based diffusion barriers (Ta-TaN layer stacks and Ta-Si-N single layers) before and after annealing experiments to elucidate the Cu diffusion process and, thus, to compare their failure mechanisms.

In the following section 2, the experimental details are introduced, while section 3 deals with the characterization of the microstructure and some functional properties of the as-deposited diffusion barriers. To better interpret the measurements for the Ta-TaN layer stacks (TaN/Ta and Ta/TaN/Ta), pure Ta and almost stoichiometric TaN single layers are primarily discussed. In section 4, the thermal stability for the practically relevant case of Ta-based barriers deposited between copper and dielectric material (SiO_2) is investigated. Besides microstructure changes, various interdiffusion processes are discussed. Sections 5 and 6 are devoted to the sensitive proof of Cu diffusion through the diffusion barriers. Besides measurements at samples with alternate film stacks, experiments employing trace-analytical techniques are introduced. Based on the experimental results, the thermal stabilities and the failure mechanisms of the various Ta-based diffusion barriers are comparatively discussed. This chapter closes with concluding remarks.

Chapter 2

EXPERIMENTAL DETAILS

2.1. Sample Preparation

Using a cluster tool (Surface Technology Systems Ltd.) equipped with load-lock, dealer, ICP (inductively coupled plasma) soft-etch chamber, Cu-PVD module, and Ta/Ta$_5$Si$_3$-PVD module, Ta-based diffusion barriers and Cu metallization layers were deposited onto blanket and thermally oxidized (100) Si wafers. In the latter case, a 140 nm thick silicon oxide layer was prepared by annealing of blanket Si wafers at $T = 1000$ °C for $t = 2$ h in oxygen ambience. For some samples, the deposition of a 100 nm thick silicon oxide film was carried out using a PECVD process based on tetraethoxyorthosilicate (TEOS, SiO$_4$C$_8$H$_{20}$) as precursor. After performing a RCA clean developed by the Radio Corporation of America, the substrates were loaded into the load-lock. To remove still existing contamination from the wafer surface, an ICP soft-etch was carried out using Ar$^+$ ions with a resulting energy of approximately 400 eV. Diffusion barrier deposition was done by so-called long-throw sputtering with an enlarged distance of 125 mm between substrate and Ta/Ta$_5$Si$_3$ target. After adjusting a base pressure of $p_0 < 2*10^{-5}$ Pa in the deposition chamber, radio frequency magnetron sputtering was performed at room temperature ($T = 25 ... 28$ °C), 1 kW forward power, and a constant Ar flow rate of $\phi_{Ar} = 5$ sccm. For the preparation of Ta-N layers, a Ta target (purity: 99.95 %) was used. Pure Ta films were deposited at $\phi_N = 0$ sccm, while for almost stoichiometric TaN layers, the N$_2$ flow rate was adjusted to $\phi_N = 3.5$ sccm, resulting in process pressures $p = 0.15$ Pa and 0.24 Pa, respectively. Using a Ta$_5$Si$_3$ target (composite material, purity: 99.5 %), reactive magnetron sputtering of Ta-Si-N layers was carried out at N$_2$ flow rates between

$\phi_N = 0$ sccm and 4 sccm leading to process pressures between $p = 0.22$ Pa and 0.30 Pa. To complete the metallization system, the wafers were transferred into the Cu-PVD module (base pressure $p_0 < 2*10^{-5}$ Pa) without interrupting the vacuum. There, direct current magnetron sputtering of a Cu layer was done at room temperature, 1 kW forward power, and a constant Ar flow rate of $\phi_{Ar} = 3$ sccm. For further investigations, the samples were cut into pieces of about 20 x 30 mm². Table 1 provides the layer stacks for all prepared samples. The given barrier compositions were determined by Rutherford backscattering spectrometry (RBS). For these experiments, the Ta-based films were deposited onto amorphous carbon substrates which allowed for an unambiguous evaluation of the Si signal in the barrier layer and N detection with higher sensitivity.

Table 1. Layer stacks of all investigated samples. The numbers put in parentheses are the film thicknesses in nanometers

Passivation	Metallization	Barrier	Substrate
		Ta (50)	Si
		TaN (50)	Si
		$Ta_{73}Si_{27}$ (10)	Si
		$Ta_{56}Si_{19}N_{25}$ (10)	Si
		$Ta_{30}Si_{18}N_{52}$ (10)	Si
	Cu (50)	Ta (10)	Si
	Cu (50)	TaN (10)	Si
	Cu (50)	Ta (10) / TaN (10) / Ta (10)	Si
	Cu (50)	$Ta_{73}Si_{27}$ (10)	Si
	Cu (50)	$Ta_{56}Si_{19}N_{25}$ (10)	Si
	Cu (50)	$Ta_{30}Si_{18}N_{52}$ (10)	Si
		Ta (50)	SiO_2 (140) / Si
		TaN (50)	SiO_2 (140) / Si
		$Ta_{73}Si_{27}$ (10)	SiO_2 (140) / Si
		$Ta_{56}Si_{19}N_{25}$ (10)	SiO_2 (140) / Si
		$Ta_{30}Si_{18}N_{52}$ (10)	SiO_2 (140) / Si
	Cu (50)	Ta (10)	SiO_2 (140) / Si
	Cu (50)	TaN (10)	SiO_2 (140) / Si
	Cu (50)	TaN (10) / Ta (10)	SiO_2 (140) / Si
	Cu (50)	Ta (10) / TaN (10) / Ta (10)	SiO_2 (140) / Si
	Cu (50)	$Ta_{73}Si_{27}$ (10)	SiO_2 (140) / Si
	Cu (50)	$Ta_{62}Si_{20}N_{18}$ (10)	SiO_2 (140) / Si

Table 1. (Continued)

Passivation	Metallization	Barrier	Substrate
	Cu (50)	$Ta_{56}Si_{19}N_{25}$ (10)	SiO_2 (140) / Si
	Cu (50)	$Ta_{41}Si_{20}N_{39}$ (10)	SiO_2 (140) / Si
	Cu (50)	$Ta_{30}Si_{18}N_{52}$ (10)	SiO_2 (140) / Si
$Ta_{56}Si_{19}N_{25}$ (50)	Cu (5)	$Ta_{56}Si_{19}N_{25}$ (100)	SiO_2 (140) / Si

2.2. Annealing Experiments

To characterize the effectiveness of the barrier layers against Cu diffusion, systematic annealing experiments were carried out. Detailed analytic investigations were performed prior and after these heat treatments to analyze changes in microstructure, layer setup, and particular functional properties. Based on this approach, structure and performance alterations of the diffusion barriers as they might occur during microprocessor manufacturing can be monitored. Furthermore, annealing experiments accelerate barrier degradation processes and thus allow for their detailed investigation.

Annealing of the barrier systems was performed using a tube furnace. To avoid the impact of a special ambience, all heat treatments were carried out under vacuum conditions with a residual gas pressure of $p_0 \approx 10^{-4}$ Pa. Two different approaches were applied. Firstly, samples were annealed at several temperatures for a fixed time of $t = 1$ h which allows for the determination of the critical temperatures for particular degradation processes. Secondly, heat treatments were done at constant temperatures (e.g. $T = 600$ °C) and varying annealing time ($t = 1 \ldots 100$ h) to analyze diffusion processes in more detail. It should be mentioned that for each annealing experiments, a dedicated sample was used.

2.3. Sample Analysis

In this study, the characterization of the diffusion barriers in the as-deposited state and after thermal stressing was carried out using X-ray scattering, spectroscopic, and microscopic techniques. In particular, these complementary analytical techniques allow for a comparable determination of the barrier stabilities and a detailed analysis of the barrier failure modes.

The characterization of the layer structure was performed using *X-ray reflectometry* (XRR) as well as *X-ray diffraction* (XRD) at a small incidence angle or in Bragg-Brentano geometry (θ-2θ scans). For layer stacks, XRR experiments provide information about the film thicknesses, the interface roughnesses, and the depth profile of the electron density. Besides the determination of the degree of crystallinity, XRD investigations are suitable to analyze phase composition, lattice parameters for each crystalline phase, crystallite size, texture, and lattice strains. In this study, the measurements were done using Cu-Kα radiation (λ = 1.5418 Å) and a Philips X'Pert Pro diffractometer equipped with X-ray mirror, Eulerian cradle, collimator, and graphite monochromator. For XRR measurements, a 0.1 mm wide slit was placed between the collimator and the graphite monochromator to enhance the angular resolution in 2θ. Additionally, a 0.125 mm thick Ni foil was automatically introduced into the primary beam to attenuate intensities higher than a critical level. Fitting of the measured XRR curves was carried out using the software IMD 4.1 [209]. The accuracies of the mass densities ρ_f and layer thicknesses h_f determined in the fitting process are given as $\Delta\rho_f = \pm 0.1$ gcm^{-3} and $\Delta h_f = \pm 0.1$ nm, respectively. The error of the Cu surface roughness is $\Delta\sigma_{vacuum/Cu} = \pm 0.3$ nm, while for all other interface roughnesses it is given as $\Delta\sigma_{layer\,A/layer\,B} = \Delta\sigma_{layer/substrate} = \pm 0.1$ nm. For sensitive phase analysis of the Ta-based diffusion barrier layers, glancing angle XRD measurements were performed at incidence angle $\omega = 2°$. Phase identification was done by comparison of the positions of the measured diffraction maxima with peak positions of potential Ta-based phases contained in the PDF (powder diffraction file) data base [210].

Glow discharge optical emission spectroscopy (GDOES) depth profiles were recorded using a LECO GDS 750 spectrometer and a radio frequency generator [211]. By simultaneous acquisition of photons emitted by various elements in an Ar plasma, a broad spectrum of the periodic table can be covered. In this study, depth profiles were obtained for copper, tantalum, silicon, nitrogen, oxygen, and carbon with a depth resolution of < 5 nm at the sample surface. The following emission lines were used: λ_{Cu} = 327.4 nm, λ_{Ta} = 362.6 nm, λ_{Si} = 288.1 nm, λ_N = 174.2 nm, λ_O = 130.2 nm, and λ_C = 156.1 nm. With acquisition times < 1 s, GDOES allows for a fast characterization of the layer setup over macroscopic sample areas (diameter of the sputter crater ≈ 2.5 mm). However, since the course of the depth profiles can be significantly influenced if conductive and nonconductive parts of the sample are present at the same time, physical effects have to be thoroughly separated form sample properties [212].

To detect Cu traces within the Ta-based diffusion barrier layers, *secondary ion mass spectroscopy* (SIMS) measurements were performed with O_2^+ primary ions of 1 keV energy employing a dynamically double-focusing secondary ion mass spectrometer IMS 6f (Cameca). The lateral extension of the sputter crater had a diameter of 150 µm, while the ^{65}Cu signal was recorded with a mass resolving power of 300 from a circular sample area of 100 µm diameter in the center of the sputter crater.

The evaluation of the diffusion barrier effectiveness necessitates the determination of the Cu contents in the barrier layer, the dielectric, and the Si substrate after thermal treatment. For this purpose, chemical-analytical methods for a selective wet-chemical etching of the different materials as well as for the evaluation of the Cu concentration in each etch solution were developed. All analyses started with etching away the Cu cap layer using 5 % ammonium persulfate solution. In a first experiment, the barrier layer and SiO_2 were dissolved together by 10 % hydrofluoric acid and analyzed by *graphite furnace atomic absorption spectrometry* (GFAAS) using an AAS6 vario tool (Analytik Jena). To determine the Cu content only within the SiO_2 film, the barrier layer was removed by etching with 48 % HBF_4 solution. Afterwards, the dielectric was dissolved in 10 % hydrofluoric acid and analyzed by GF-AAS.

Transmission electron microscopy (TEM) offers the advantage to carry out microstructure investigations at cross-sections of layer stacks with high lateral resolution [213]. In particular, TEM bright field images are suitable for a representation of the layer setup and for a determination of the layer thicknesses. Additionally, conclusion about the crystallinity of the barrier layer can be drawn by lattice plane imaging. In the case of crystalline films, phase analyses can be performed using electron diffraction. Analytical TEM including *energy dispersive X-ray spectroscopy* (EDXS) and *electron energy loss spectroscopy* (EELS) allows for the determination of the chemical composition in small sample volumes. In this study, TEM investigations were carried out employing a FEI Tecnai F30 transmission electron microscope equipped with EDX detector (EDAX) and electron energy loss spectrometer (Gatan). Cross-sectional sample preparation was done by grinding, dimpling, and Ar ion milling, which leads to a final specimen thickness between 20 and 40 nm.

Chapter 3

MICROSTRUCTURE AND FUNCTIONAL PROPERTIES OF AS-DEPOSITED TA-BASED DIFFUSION BARRIERS

As shown in the literature review, the performance of a diffusion barrier depends significantly on its chemical composition as well as on its microstructure. It is the latter one which not only determines the reaction behavior with the dielectric, but also the interdiffusion of various elements and, consequently, that of copper. Furthermore, the barrier microstructure has a significant impact on mechanical, thermal, and electrical properties, such as the residual film stress, the coefficient of thermal expansion, or the resistivity. In addition, the microstructure of a diffusion barrier can influence that of the Cu layer. As a consequence, a detailed microstructure analysis regarding to phase composition, grain size, and preferred grain orientations is necessary for all as-deposited diffusion barriers.

3.1. Ta-TaN Layer Stacks

In the as-deposited state, a *Ta diffusion barrier* deposited onto thermal oxide or directly onto silicon predominantly consists of nanocrystalline tetragonal β-Ta (Figure 3). For a nominal 10 nm thick layer, the Ta crystallite size evaluated by Scherrer's equation [214] (d_{Ta} = 9.4 ± 2.3 nm) is comparable to the layer thickness determined by XRR measurements (h_{Ta} = 10.4 nm). Thus, direct Cu diffusion along Ta grain boundaries into the substrate is expected at elevated temperatures. Since θ-2θ scans at increasing amount of the tilt angle ψ result in a shift of the

002 β-Ta peak position ($2\theta \approx 33.6°$) to larger diffraction angles, it can be concluded that the Ta layer is characterized by a residual compressive stress after deposition. Furthermore, plotting the maximum 002 β-Ta peak intensity over the tilt angle ψ shows that the β-Ta crystallites are preferentially oriented with the c-axis of the unit cell perpendicular to the sample surface. Besides β-Ta Bragg reflections, the diffraction pattern in Figure 3 is characterized by additional peaks which can be assigned to polycrystalline copper. Due to the used diffraction geometry with incidence angle $\omega = 2°$, a strong Cu 220 reflection at $2\theta = 74.2°$ points to the presence of preferentially <111>-oriented Cu grains.

Besides a small amorphous fraction, a *TaN diffusion barrier* deposited onto blanket or thermally oxidized silicon consists mainly of nanocrystalline fcc TaN (Figure 3). Contrary to tantalum, the grain size ($d_{TaN} = 2.9 \pm 0.6$ nm) is much smaller than the film thickness ($h_{Ta} = 14.7$ nm). TaN layers are characterized by a compressive residual stress. Since recording the diffracted intensity of the 111 TaN Bragg reflection depending on the tilt angle ψ leads to maxima at $\psi = 0°$ and $\psi \approx \pm 70.5°$, the presence of a <111> texture component is concluded.

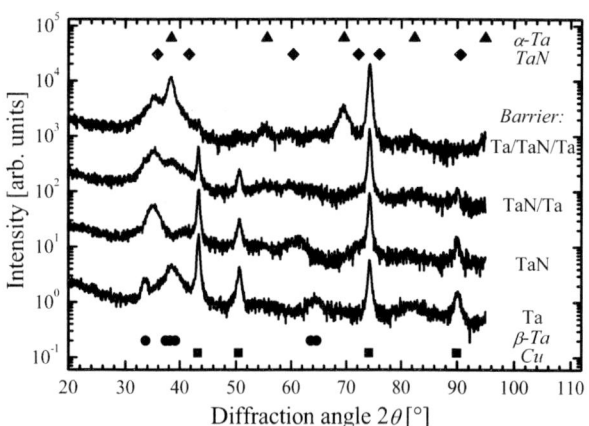

Figure 3. Glancing angle XRD diagrams of the samples Cu/Ta/SiO$_2$/Si, Cu/TaN/SiO$_2$/Si, Cu/TaN/Ta/SiO$_2$/Si, and Cu/Ta/TaN/Ta/SiO$_2$/Si in the as-deposited state.

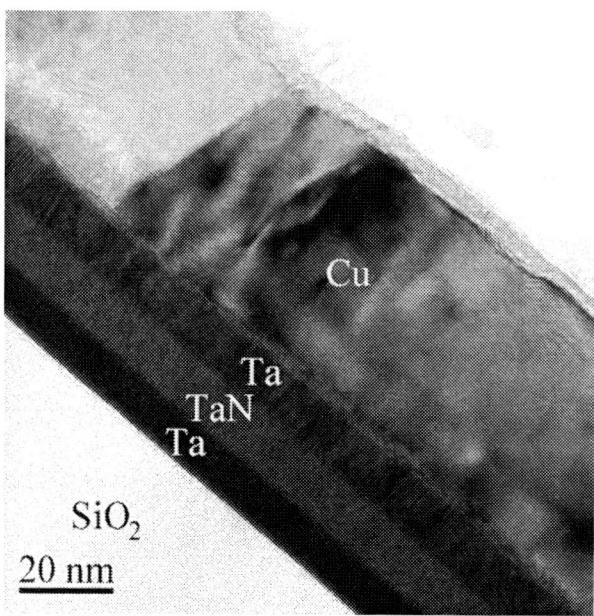

Figure 4. Cross-sectional TEM bright field image of the Cu/Ta/TaN/Ta/SiO$_2$/Si sample in the as-deposited state.

Regarding the *TaN/Ta diffusion barrier*, the diffraction pattern for the Cu/TaN/Ta/SiO$_2$/Si sample in Figure 3 can be described as a superposition of the XRD diagrams for the Cu/Ta/SiO$_2$/Si sample (nanocrystalline tetragonal β-Ta) and the Cu/TaN/SiO$_2$/Si sample (nanocrystalline fcc TaN). Due to X-ray absorption in the upper TaN layer, the intensity of the β-Ta Bragg reflections is, however, significantly reduced. As for the bilayer system, glancing angle XRD diagrams for the *Ta/TaN/Ta diffusion barrier* show fcc TaN diffraction maxima and small signs of tetragonal β-Ta (Figure 3). In addition, predominantly <110>-oriented α-Ta is observed. According to the investigations done so far, it can be concluded that the α-Ta is located in the upper Ta film, whereas the lower one consists of β-Ta and the layer in between of fcc TaN. The preferred growth of low-resistivity α-Ta on TaN is in good accordance with literature results [185, 187, 215]. Due to sharp interfaces, the TEM bright field image in Figure 4 clearly shows the threefold gradation of the barrier layer. Furthermore, the crystallite sizes in the upper and lower Ta layer are in the order of the film thicknesses, while the TaN grains in the middle layer are characterized by a mean size of $d_{TaN} \approx 3$ nm which confirms the XRD results.

3.2. Ta-Si-N Single Layers

Figure 5 shows the chemical compositions of the *Ta-Si-N diffusion barriers*. For all layers, the Si content is almost constant at $x_{Si} \approx 20$ at.%, while the N content increases linearly with increasing N_2 flow rate during deposition. Correspondingly, the Ta content in the Ta-Si-N layers decreases. RBS experiments also detected small Ar and O contaminations. According to auger electron spectroscopy (AES) measurements, the O content is smaller than 2 at.%.

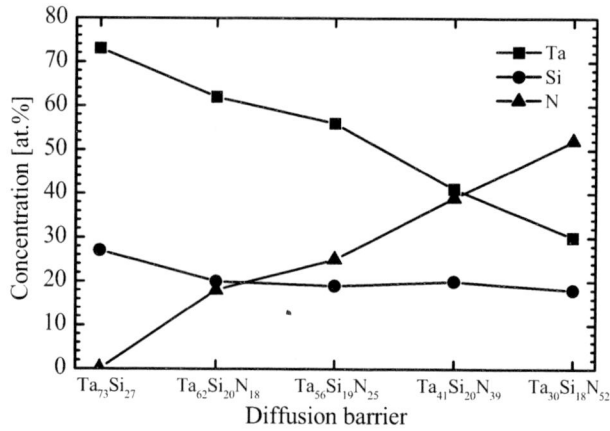

Figure 5. Chemical composition of the as-deposited Ta-Si-N layers determined by RBS.

To characterize the morphology of the Ta-Si-N layers deposited between copper and thermal oxide, XRR experiments were carried out (Figure 6). Due to a constant chemical composition of the Cu layer, the critical angle for total reflection is the same for all samples ($\theta_{c,Cu} = 0.39°$, Figure 6) and corresponds to a mass density of $\rho_{Cu} = 8.8$ gcm^{-3}. All Ta-Si-N layers are characterized by a higher mass density. With increasing N content in the barrier, $\rho_{Ta-Si-N}$ decreases from $\rho_{Ta_{73}Si_{27}} = 14.1$ gcm^{-3} to $\rho_{Ta_{30}Si_{18}N_{52}} = 10.0$ gcm^{-3}. The oscillations in the XRR curves are associated with the Cu film and the barrier layer. They allow for the determination of the corresponding film thicknesses as well as the evaluation of the interface roughnesses. Due to the polycrystalline Cu microstructure, the surface roughness is relatively high ($\sigma_{vacuum/Cu} \approx 2.4$ nm) which can be concluded from the strong damping of the high-frequency oscillations for $\theta > 0.7°$ (Figure 6). In contrast to that, the low-frequency barrier oscillations are also observed for significantly larger scattering angles which points to smooth and sharp Cu/barrier interfaces ($\sigma_{Cu/barrier} = 0.1 \ldots 0.3$ nm).

According to glancing angle XRD experiments, all Ta-Si-N layers are characterized by an amorphous microstructure which is confirmed by electron diffraction at cross-sectional TEM samples. More detailed information about the bonding structure in the Ta-Si-N layers can be obtained by X-ray photoelectron spectroscopy (XPS) investigations [216]. According to Nötzold [217] and Fischer [218], Si atoms are bonded to Ta atoms in the case of a $Ta_{73}Si_{27}$ layer. With increasing N content in the barrier layer, the fraction of Ta-N bonds increases. For a $Ta_{41}Si_{20}N_{39}$ film, there are first signs for Si-N bonds which can be unambiguously observed for a $Ta_{30}Si_{18}N_{52}$ layer [217, 218]. Due to the amorphous microstructure, the residual stress in the Ta-Si-N layers could be only determined by the substrate curvature method [219]. For 50 nm thick films, the amount of the intrinsic compressive stress varies between $|\sigma|_{min} = 300$ MPa and $|\sigma|_{max} = 820$ MPa. As in the case of the Ta-TaN layer stacks, the Cu film is characterized by a polycrystalline microstructure. The Cu crystallites are preferentially <111>-oriented and the volume fraction of this texture component decreases with increasing N content in the Ta-Si-N barrier.

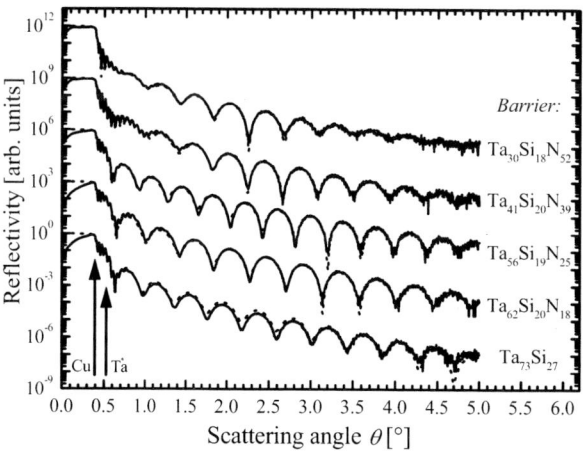

Figure 6. XRR curves for the as-deposited Cu/Ta-Si-N/SiO$_2$/Si samples (solid lines: experimental data, dotted lines: fit results). As an orientation, the critical angles for total reflection in the case of a Cu layer and a pure Ta layer are given additionally.

Figure 7 shows the electrical resistivities ρ determined for 10 nm thick Ta-Si-N barriers with the four-point probe method. For a $Ta_{73}Si_{27}$ layer as well as the N-poor $Ta_{62}Si_{20}N_{18}$ and $Ta_{56}Si_{19}N_{25}$ layers, the resistivity is small and almost constant ($\rho \approx 240$ $\mu\Omega$ cm), while there is an increase in ρ for barriers containing higher N contents. A high absolute value $\rho = 1900$ $\mu\Omega$ cm for a $Ta_{30}Si_{18}N_{52}$ layer and strong resistivity changes for slight variations of the $Ta_{41}Si_{20}N_{39}$ composition might turn out as critical issues for an industrial application of these two barriers.

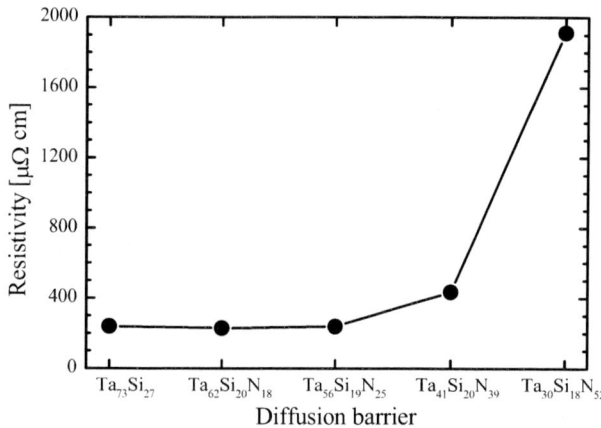

Figure 7. Resistivities of the investigated Ta-Si-N diffusion barriers.

Chapter 4

THERMAL STABILITY OF TA-BASED DIFFUSION BARRIERS ON SiO_2

For characterizing the thermal stability, which is one of the most important barrier properties (cf. section 1.1), and for studying the degradation mechanisms, a targeted initiating of the barrier failure is necessary. To distinguish between processes starting at the Cu/barrier interface and at the barrier/SiO_2 interface, respectively, annealing experiments for samples without metallization layer are very helpful. Based on the findings of such analyses, the influence of the Cu film on the microstructure changes during heat treatment can be discussed more thoroughly.

4.1. Ta-TaN Layer Stacks

4.1.1. Reaction Behavior with SiO_2

As revealed by XRD measurements, which are summarized in Table 2 and are discussed in more detail in [220], annealing of a *Ta layer* at $T = 500$ °C leads to an increase of the intrinsic compressive stress. An equivalent result was obtained by Cabral et al. [221] who carried out repeated stress temperature measurements up to $T = 400$ °C for a 180 nm thick β-Ta layer. During each thermal cycle, a compressive stress increase of $|\sigma| = 0.5$ GPa was observed and explained by O incorporation into the Ta film [221]. Principally, there are two possible O sources: (*i*) the residual gas of the annealing ambience and (*ii*) the native oxide layer on the Ta surface whose thickness was determined to $h_{Ta_2O_5}$ = 2.8 nm by XRR experiments. According to Samsonova [222] and Powers et al. [223], the coefficients for O diffusion in Ta_2O_5 and tantalum at T = 500 °C are given by D_{O/Ta_2O_5} = 3.2*10^{-14} cm^{2}s^{-1} and $D_{O/Ta}$ = 2.8*10^{-10} cm^{2}s^{-1}, respectively. The corresponding mean diffusion distances $l = (Dt)^{1/2}$ can be estimated with t = 1 h to l_{O/Ta_2O_5} = 110 nm and $l_{O/Ta}$ = 10 µm. As a consequence, O atoms, which are present to a small amount in the annealing ambience, can penetrate the Ta oxide layer and get incorporated into the Ta film. A second source of oxygen is the native Ta oxide film itself. According to Cros et al. [224], its decomposition starts at T = 550 °C allowing liberated O atoms to diffuse into underlying tantalum. As shown by Giber et al. [225], the dissolution of a 100 nm thick Ta_2O_5 layer on polycrystalline tantalum can already occur during annealing at T > 400 °C in ultrahigh vacuum. Due to reduced O supply, a smaller increase of the compressive stress during annealing is expected for Ta layers capped with copper. On the one hand, the Cu film is deposited without interrupting the vacuum which avoids the formation of native Ta oxide. On the other hand, the Cu cap layer reduces the intensity of O diffusion from the annealing ambience into the Ta barrier.

For heat treatment at $T \geq$ 600 °C, strong diffusion of C atoms into the Ta layer is observed [220]. Based on the diffusion coefficient $D_{C/Ta}$ = 2.18*10^{-9} cm^{2}s^{-1} [226], the mean diffusion distance for t = 1 h is given by $l_{C/Ta}$ = 28 µm. Since additional annealing experiments in Ar flow or at improved vacuum conditions ($p_0 \approx 10^{-5}$ Pa) result in significantly reduced C diffusion, hydrocarbons contained in the residual gas of the annealing ambience ($p_0 \approx 10^{-4}$ Pa) seem to be the C source. According to XRD investigations, C diffusion into tantalum leads to the formation of hexagonal Ta_2C (Table 2). Although both processes can significantly worsen the electrical and mechanical properties of a Ta diffusion barrier, the incorporation of carbon seems to improve its thermal stability [227].

Besides the commencing Ta_2C formation, additional Bragg reflections are observed in the diffraction pattern after annealing at T = 600 °C. These diffraction maxima cannot be explained by known Ta compounds or a change of the preferred orientation of the β-Ta crystallites. Assuming, however, a second phase (β'-Ta) with tetragonal unit cell and lattice parameters which are slightly larger (a = 10.44(5) Å, c = 5.58(2) Å) than those of β-Ta, a satisfactory description is possible (Table 2). Since annealing in Ar flow (with reduced C content) does not influence the additional Bragg reflections, the diffusion of C atoms into the Ta film seems to be not responsible for the formation of the β'-Ta phase. However, its stabilization by impurity atoms cannot be completely excluded. As shown above, even under UHV conditions, the decomposition of the native Ta oxide film leads to diffusion of liberated O atoms into the Ta layer. Heat treatment up to $T = 700$ °C results in a growth of the β'-Ta phase at the expense of β-Ta. At $T = 750$ °C, both phases transform completely into thermodynamically stable α-Ta (Table 2). According to Clevenger et al. [145], the abrupt transformation of β-Ta into α-Ta is observed for 100 nm thick Ta films also at $T = 750$ °C. The transformation process is believed to be the main stress relief mechanism in the Ta films that are intrinsically compressively stressed [145]. In combination with GDOES and TEM results, glancing angle XRD investigations indicate that bcc α-Ta reacts with thermal oxide to orthorhombic Ta_2O_5 and tetragonal Ta_5Si_3 at $T = 1000$ °C (Table 2) which is in accordance with theoretical considerations reported by Pretorius et al. [228] and Beyers [229]. In general, the Ta silicide grains grow on the remaining SiO_2 substrate, while the Ta oxide is detected at the sample surface. It should be mentioned that the Ta_5Si_3 formation cannot be explained by diffusion of Si atoms from the Si substrate through the thermal oxide. Based on the coefficient for Si diffusion in amorphous SiO_2 at $T = 1100$ °C ($D_{Si/SiO_2} = 3.08*10^{-20}$ cm^{2}s^{-1} [230]), the mean diffusion distance is calculated to $l_{Si/SiO_2} = 1.1$ Å which is much smaller than the oxide thickness.

Table 2. Phase formation behavior for the Ta/SiO$_2$/Si and the TaN/SiO$_2$/Si system during annealing at temperatures between 500 °C ≤ T ≤ 1100 °C for $t = 1$ h

	Ta/SiO$_2$/Si	TaN/SiO$_2$/Si
as-depos.	β-Ta	TaN, amorphous
500°C / 1h	β-Ta	TaN, amorphous
550°C / 1h	β-Ta, β'-Ta	-
600°C / 1h	β-Ta, β'-Ta, Ta$_2$C	TaN, amorphous
650°C / 1h	β-Ta, β'-Ta, Ta$_2$C	-
700°C / 1h	β-Ta, β'-Ta, Ta$_2$C	TaN, amorphous
750°C / 1h	Ta$_2$C, α-Ta	-
800°C / 1h	Ta$_2$C, α-Ta	TaN, amorphous
900°C / 1h	Ta$_2$C, α-Ta, Ta$_2$O$_5$	TaN, amorphous
1000°C / 1h	Ta$_2$C, Ta$_2$O$_5$, Ta$_5$Si$_3$	Ta$_4$N$_5$

Table 2. (Continued)

	Ta/SiO$_2$/Si	TaN/SiO$_2$/Si
1050°C / 1h	-	Ta$_4$N$_5$, Ta$_5$N$_6$
1100°C / 1h	Ta$_2$C, Ta$_2$O$_5$, Ta$_5$Si$_3$	Ta$_4$N$_5$, Ta$_5$N$_6$

Samples belonging to boxes marked with an "-" were not investigated.

Heat treatment of a *TaN layer* up to $T = 800$ °C leads not only to a decrease of the amorphous phase fraction and a decrease of the cubic lattice parameter, but also to TaN crystallite growth. In particular, the mean TaN grain size for a 50 nm thick layer increases from $d_{TaN} = 3.5$ nm in the as-deposited state to $d_{TaN} = 10$ nm after annealing at $T = 900$ °C (Figure 8). Although there is an increase of the diffracted intensity for $2\theta \approx 26°$ already at $T = 800$ °C, Bragg reflections of tetragonal Ta$_4$N$_5$ are identified at this position only after annealing at $T = 1000$ °C. The beginning of Ta$_4$N$_5$ formation can be explained by a disordered growth of TaN crystallites. Due to $x_N/x_{Ta} = 1.1$ in the as-deposited films and, thus, a small excess supply of nitrogen, the Ta sub-lattice of fcc TaN cannot completely occupied. This results in a decrease of the bond lengths and, consequently, a decrease of the TaN lattice parameter. The continuous approach to the Ta$_4$N$_5$ structure, which can be described as a tetragonally distorted fcc TaN lattice due to an incomplete occupation of the Ta sub-lattice [170, 231], is characterized by the above-mentioned intensity increase at small diffraction angles. At sufficient thermal activation ($T = 1000$ °C), the complete transformation from fcc TaN into tetragonal Ta$_4$N$_5$ occurs. Further heat supply leads to the formation of hexagonal Ta$_5$N$_6$. There are, however, no signs for a reaction with SiO$_2$ up to $T = 1100$ °C.

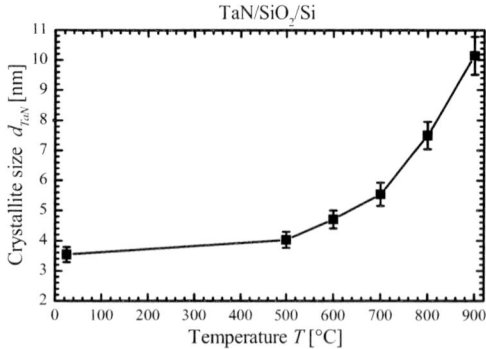

Figure 8. Dependency of the TaN crystallite size d_{TaN} for the TaN/SiO$_2$/Si sample on the annealing temperature T.

4.1.2. Structure Changes for Cu-Capped Ta-TaN Layer Stacks

According to the previous subsection, the thermal stability of Ta and TaN layers on thermal oxide is sufficient for their application as diffusion barriers in Cu metallization systems. In particular, the direct comparison between both layers showed that N addition during deposition leads to a decreased propensity for a chemical reaction with the dielectric material. The effects of a Cu metallization layer on the annealing behavior are discussed in this subsection. For further details, the reader is referred to [232].

For a *Ta diffusion barrier* between copper and SiO_2, there is no significant stress increase up to $T = 500$ °C. Thus, the presence of the Cu layer seems to hinder O incorporation into tantalum. However, an increased Ta signal is detected by GDOES at the Cu surface (Figure 9 (a), curve (3)). It can be explained by commencing diffusion of Ta atoms through the Cu metallization layer. Such Ta diffusion, which leads to a shift of the Cu profile to longer sputter times (Figure 9 (b), curve (4)), was also observed by Jang et al. [161]. At $T = 700$ °C, tetragonal β-Ta transforms into bcc α-Ta which shows a preferred <110> orientation (Table 3). According to TEM analyses, the α-Ta crystallites grow on the SiO_2 substrate and within the original Ta layer. Thus, the Ta signal in the Cu region (Figure 9 (a), curve (5)) is caused by intensified Ta diffusion. As a result of all changes in the Ta layer, copper gets into contact with the substrate (Figure 9 (b), curve (5)) and finally diffuses into SiO_2 which was confirmed by EELS line scans.

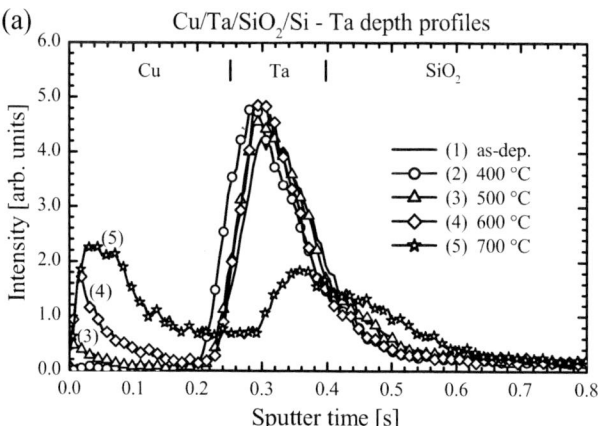

Figure 9 (Continued)

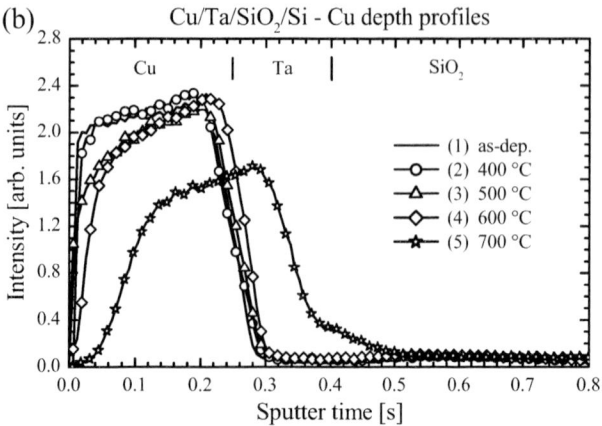

Figure 9 (a) reprinted from reference [232] with permission from Elsevier.

Figure 9. GDOES depth profiles of the Ta distribution (a) and of the Cu distribution (b) for the Cu/Ta/SiO$_2$/Si sample in the as-deposited state and after annealing at several temperatures T for t = 1 h.

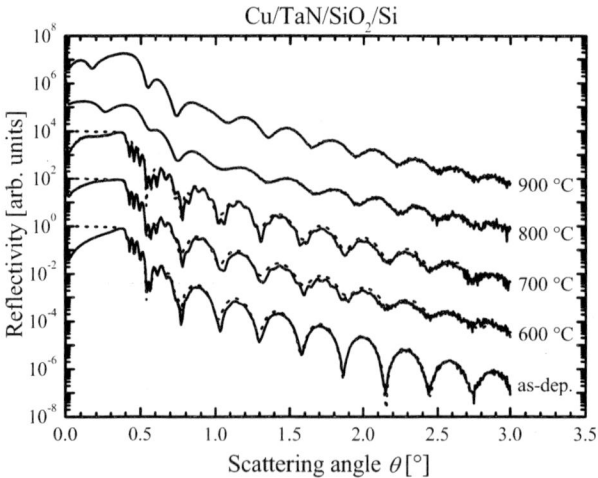

Figure 10. XRR curves for the Cu/TaN/SiO$_2$/Si sample in the as-deposited state and after annealing at several temperatures T for t = 1 h (solid lines: experimental data, dotted lines: fit results).

Since glancing angle XRD diagrams show the same changes for the TaN Bragg reflections as in the case of the TaN/SiO$_2$/Si sample, the presence of a Cu layer has principally no influence on the microstructure evolution of a *TaN diffusion barrier*. Concerning the Cu layer, small intensity changes are observed only for $T > 700\ °C$. Contrary to the XRD patterns, the XRR curves undergo stronger changes during heat treatment. At $T = 600\ °C$, the amplitude of the barrier oscillations decreases (Figure 10) which points to a slight increase of the Cu/TaN interface roughness. According to GDOES measurements, this effect seems to be caused by O diffusion from the residual gas of the annealing ambience through the Cu layer. Ta diffusion out of the TaN barrier cannot account for the roughness increase (Figure 11 (a), curve (2)).

For $T \geq 800\ °C$, additional changes are observed in the XRR curves. Besides an intensity dip at small scattering angles, the Cu layer oscillations are no longer visible (Figure 10). In the GDOES depth profiles, the TaN barrier signal appears at the sample surface and copper is detected in the substrate region (Figures 11 (a) and (b), curves (4), (5)). On a first glance, these observations point to strong Cu diffusion into SiO_2. However, this explanation would be inconsistent with the XRD results which show Bragg reflections of polycrystalline copper. A detailed scanning electron microscopy (SEM) analysis finally reveals that copper agglomerates and forms islands at $T \geq 800\ °C$ (Figure 12). Additional AES measurements between the Cu islands exclude strong Cu diffusion into the TaN barrier.

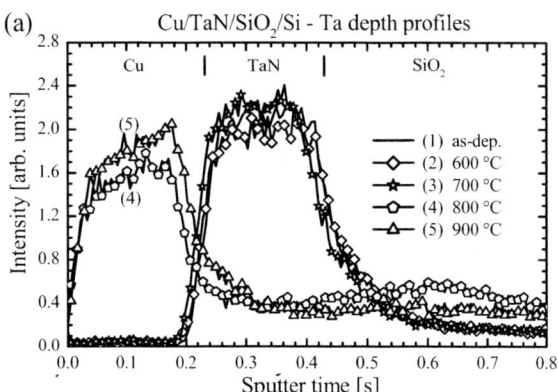

Figure 11 (Continued)

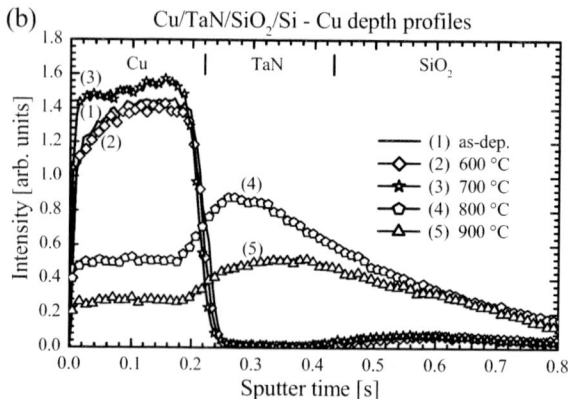

Figure 11. GDOES depth profiles of the Ta distribution (a) and of the Cu distribution (b) for the Cu/TaN/SiO$_2$/Si sample in the as-deposited state and after annealing at several temperatures T for $t = 1$ h.

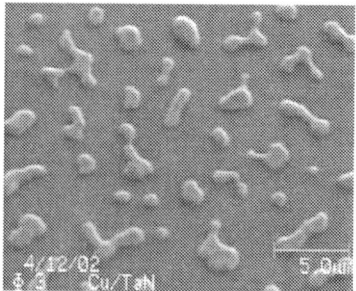

Figure 12. SEM surface image of the Cu/TaN/SiO$_2$/Si sample after annealing at $T = 900$ °C for $t = 1$ h.

The formation of islands is based on the fact that a thin layer deposited onto a substrate is characterized by a large surface area compared to its volume. Transport of matter via diffusion processes can lead to a decrease of the surface area and, thus, of the surface energy. Additionally, diffusion leads to shape changes, to film rupture, and finally to island formation. The process is called agglomeration and occurs in two steps: (*i*) the formation of holes within the layer and (*ii*) the growth of these holes. For the first step, fluctuations of the layer thickness, which have to be large enough to penetrate the whole film, are necessary [233]. Such fluctuations can be caused by groove formation and arise at positions where grain boundaries intersect the film surface. Assuming that there is only a single grain boundary, the groove profile has a time-independent shape whose linear dimensions are proportional to $t^{1/4}$ [234]. In the case of polycrystalline films, grain boundary grooves are characterized by a finite depth which depends on the layer thickness as well as on the crystallite size [235]. If the groove intercepts the substrate, either the film will exist as contiguous islands all meeting along lines (the traces of the former grain boundaries on the substrate) or the bumps will contract to form separated islands. A quantitative analysis of the film stability based on energetic arguments is given by Srolovitz et al. [235].

In contrast to Ta and TaN single layers, significant microstructure changes occur for a *TaN/Ta diffusion barrier* at temperatures as low as $T = 300$ °C and 400 °C. The observed shifts of the β-Ta and TaN Bragg reflections in the XRD patterns can be explained by N diffusion from the TaN layer into the adjacent Ta film (Figure 13 (b), curve (2)). At $T = 500$ °C, this N diffusion results in the formation of hexagonal Ta_2N (Table 3). Both processes are intensified for higher annealing temperatures. However, at $T = 700$ °C, remnants of fcc TaN are observed in the diffraction pattern (Table 3) and the N distribution of the whole barrier is still inhomogeneous (Figure 13 (b), curve (3)). Ta diffusion out of the barrier region leads to an increased Ta surface signal (Figure 13 (a), curve (3)).

The microstructure changes observed for a *Ta/TaN/Ta diffusion barrier* are comparable to those detected for a TaN/Ta bilayer. In particular, N diffusion from the TaN layer into the adjacent Ta films starts at $T = 300$ °C and results in Ta_2N formation at $T = 500$ °C (Table 3). After annealing at $T = 600$ °C, the originally threefold graded Ta/TaN/Ta barrier has almost completely transformed into a homogeneous Ta_2N film. In contrast to the TaN/Ta barrier, diffusion of Ta atoms to the sample surface starts at $T = 500$ °C.

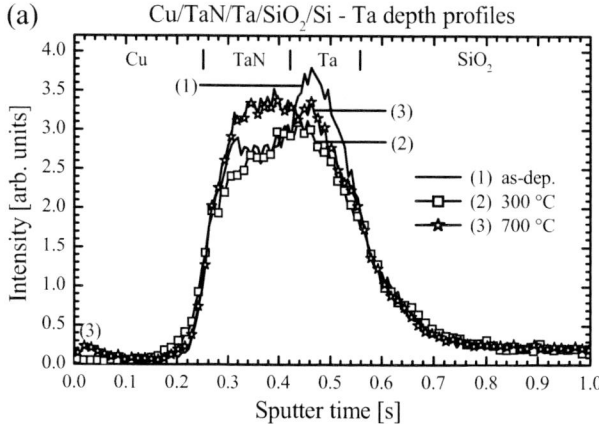

Figure 13 (Continued)

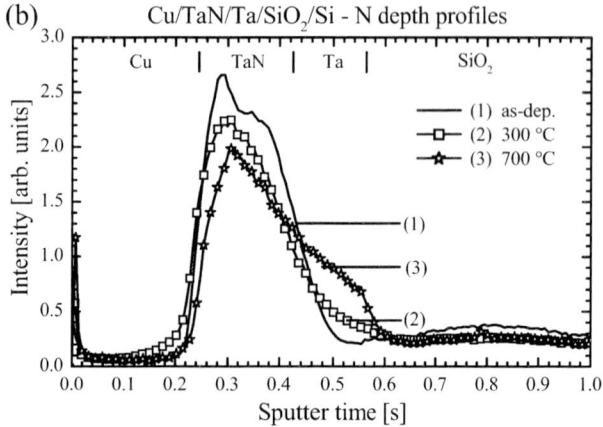

Reprinted from reference [232] with permission from Elsevier.

Figure 14. GDOES depth profiles of the Ta distribution (a) and of the N distribution (b) for the Cu/TaN/Ta/SiO$_2$/Si sample in the as-deposited state and after annealing at T = 300 °C and 700 °C for t = 1 h.

Table 3. Crystalline phase formation behavior for different Cu/barrier/SiO$_2$/Si systems during annealing at temperatures between 300 °C ≤ T ≤ 700 °C for t = 1 h

	Cu/Ta/SiO$_2$/Si	Cu/TaN/SiO$_2$/Si	Cu/TaN/Ta/SiO$_2$/Si	Cu/Ta/TaN/Ta/SiO$_2$/Si
as-depos.	Cu, β-Ta	Cu, TaN	Cu, TaN, β-Ta	Cu, α-Ta, TaN, β-Ta
300°C / 1h	Cu, β-Ta	Cu, TaN	Cu, TaN, β-Ta	Cu, α-Ta, TaN, β-Ta
400°C / 1h	Cu, β-Ta	Cu, TaN	Cu, TaN, β-Ta	Cu, α-Ta, TaN, β-Ta
500°C / 1h	Cu, β-Ta	Cu, TaN	Cu, TaN, β-Ta, Ta$_2$N	Cu, α-Ta, TaN, β-Ta, Ta$_2$N
600°C / 1h	Cu, β-Ta	Cu, TaN	Cu, Ta$_2$N, TaN	Cu, Ta$_2$N
700°C / 1h	Cu, α-Ta	Cu, TaN	Cu, Ta$_2$N, TaN	Cu, Ta$_2$N

As shown above, there are two main diffusion processes which occur for Ta-TaN layer stacks deposited between copper and silicon oxide: (*i*) Ta diffusion from the barrier to the sample surface and (*ii*) N redistribution within the barrier. In the following, both processes are discussed in more detail.

Ta diffusion through the Cu metallization layer to the sample surface is detected depending on the chemical composition of the barrier layer. For a single Ta film, this process starts at $T = 500\,°C$ (Figure 9 (a), curve (3)), while there are no signs for Ta diffusion up to $T = 700\,°C$ in the case of a TaN barrier (Figure 11 (a)). This result can be explained with a stronger interaction between Ta and N atoms compared to Ta atoms among each other which is expressed by a negative standard formation enthalpy for TaN ($\Delta^f H^0_{TaN} = -252$ kJmol^{-1} [236]). The arrangement of the Ta and TaN films in a layer stack influences the critical temperature for Ta diffusion, too. In the case of a TaN/Ta bilayer, a Ta surface signal is observed for the first time at $T = 700\,°C$ (Figure 13 (a), curve (3)), since Ta atoms have to penetrate the nanocrystalline TaN film before they can diffuse along Cu grain boundaries to the sample surface. However, for a Ta/TaN/Ta trilayer, Ta diffusion is already detected at $T = 500\,°C$. Based on the above findings, it can be concluded that the diffusing Ta atoms originate from the upper α-Ta layer. The high affinity between tantalum and oxygen ($\Delta^f G_{Ta_2O_5} = -1648$ kJmol^{-1} at $T = 627\,°C$ [236]) significantly influences the Ta diffusion process which finally leads to a decrease of the barrier layer thickness and, consequently, a reduced barrier stability against Cu diffusion. Furthermore, incorporation of Ta atoms changes the electrical properties of the Cu metallization layer. Thus, efforts must be undertaken to prevent the disadvantageous Ta diffusion.

The combination of Ta and TaN films to layer stacks can result in a second diffusion phenomenon. During annealing at temperatures as low as $T = 300\,°C$, N atoms from the TaN layer diffuse into the adjacent Ta films (Figure 13 (b), curve (2)). Since the TaN layer is characterized by $x_N/x_{Ta} = 1.1$, excess N atoms seem to initiate the N redistribution process. At $T = 500\,°C$, hexagonal Ta$_2$N is formed (Table 3). With an energy transfer of $\Delta G_{Ta_2N} = -12.7$ kJmol^{-1} at $T = 527\,°C$ [236], the reaction between Ta and TaN proceeds spontaneously. In the case of a TaN/Ta bilayer, this reaction is incomplete and a small fraction of fcc TaN remains in the barrier. For a Ta/TaN/Ta trilayer, the amount of tantalum is sufficient to transform all fcc TaN into Ta$_2$N (Table 3). Both the N redistribution and the subsequent Ta$_2$N formation degrade the barrier integrity. Additionally, both processes result in microstructure changes which not only have a negative influence on physical properties (e.g. resistivity), but also on the barrier stability. However, for a sensitive proof of Cu diffusion through the barrier, trace analysis techniques with extremely low detection limits for copper are necessary (section 5).

4.2. Ta-Si-N Single Layers

4.2.1. Reaction Behavior with SiO_2

As in the case of Ta and TaN layers, the thermal stability of ternary Ta-Si-N films with the interfacing dielectric is a necessary precondition for an application of these Ta-based materials as a diffusion barrier in Cu metallization systems. Since the X-ray amorphous microstructure remains stable for all investigated Ta-Si-N layers at annealing up to $T = 900$ °C, the required chemical stability with silicon oxide is ensured. However, to find out which mechanisms are responsible for a degradation of the layer integrity, more intensive heat treatments were performed.

For a $Ta_{73}Si_{27}$ *layer*, formation of orthorhombic Ta_2O_5 and, thus, a reaction with SiO_2 starts at $T = 950$ °C (Table 4). In the case of a $Ta_{56}Si_{19}N_{25}$ *layer*, Ta_2N nucleation occurs at $T = 1000$ °C (Table 4). Besides additional Ta_2O_5 formation at $T = 1050$ °C, more intensive heat treatment leads to the nucleation of $TaN_{0.8}$ (Table 4). In accordance with the XRD results, first significant changes of the GDOES depth profile are detected at $T = 1000$ °C (Figure 14 (a), curve (2)). Due to a further decrease of the Si signal in the $Ta_{56}Si_{19}N_{25}$ layer region (Figure 14 (a), curve (3)) at $T = 1050$ °C and almost constant Ta and N distributions, it is concluded that the Ta_2N crystallites grow within the original $Ta_{56}Si_{19}N_{25}$ layer and that silicon is not incorporated into this Ta nitride. As in the case of the $Ta_{73}Si_{27}$ film, an increased O signal in the lower part of the original Ta-based layer and a shift of the corresponding SiO_2 signal to smaller sputter times point to a reaction with the substrate (Figure 14 (b), curves (3), (4)). If Ta_2O_5 would be the result of a reaction with residual oxygen from the annealing ambience, the SiO_2 signal should be shifted to larger sputter times. A further increase of the N content in the Ta-Si-N film does not lead to principal changes of the annealing behavior. First indications for $Ta_{30}Si_{18}N_{52}$ *layer* crystallization into fcc TaN are detected at $T = 1000$ °C. The commencing reaction with SiO_2 at $T = 1100$ °C results not only in a decreased TaN phase fraction, but also in the formation of orthorhombic Ta_2O_5 (Table 4).

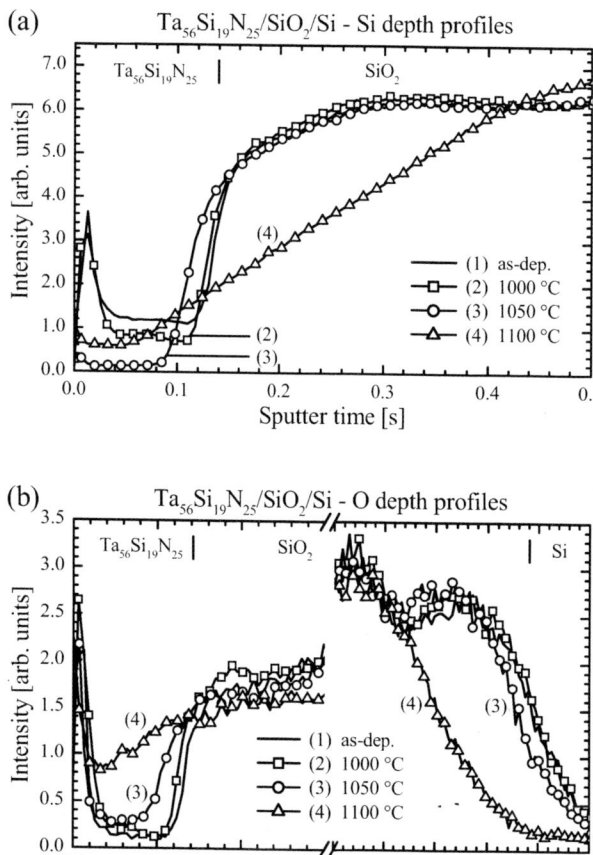

Figure 14. GDOES depth profiles of the Si distribution (a) and of the O distribution (b) for the $Ta_{56}Si_{19}N_{25}/SiO_2/Si$ sample in the as-deposited state and after annealing at several temperatures T for $t = 1$ h.

In summary, there are two processes which degrade the integrity of Ta-Si-N layers deposited onto thermal oxide: A $Ta_{73}Si_{27}$ film reacts with SiO_2 to form Ta_2O_5, while for layers containing nitrogen, a crystallization process is observed first. The stoichiometry of the crystallization product depends on the chemical composition of the Ta-Si-N film. Ta_2O_5 formation as a consequence of a reaction with SiO_2 is detected for Ta-Si-N layers, too. The corresponding reaction temperatures are, however, higher than the crystallization temperatures.

According to the XRD results, the propensity for a reaction with SiO_2 decreases with increasing N content in the Ta-Si-N layer (Table 4).

Table 4. Phase formation behavior for Ta-Si-N/SiO_2/Si systems during annealing at temperatures between 900 °C $\leq T \leq$ 1100 °C for t = 1 h

	$Ta_{73}Si_{27}$/SiO_2/Si	$Ta_{56}Si_{19}N_{25}$/SiO_2/Si	$Ta_{30}Si_{18}N_{52}$/SiO_2/Si
as-depos.	amorphous	amorphous	amorphous
900°C / 1h	amorphous	amorphous	amorphous
950°C / 1h	amorphous, Ta_2O_5	amorphous	amorphous
1000°C / 1h	amorphous, Ta_2O_5	amorphous, Ta_2N	amorphous, TaN
1050°C / 1h	Ta_2O_5	amorphous, Ta_2N, Ta_2O_5	amorphous, TaN
1100°C / 1h	Ta_2O_5	$TaN_{0.8}$, Ta_2O_5	TaN, Ta_2O_5

Table 5. Summary of the annealing times t in [h] (at T = 600 °C) necessary for the beginning of the various interdiffusion processes (determined by GDOES depth profiling) and the beginning of the barrier crystallization together with the corresponding crystallization products (determined by glancing angle XRD) for the Cu/Ta–Si–N/SiO_2/Si samples (n.o.: not observed)

Diffusion barrier	Diffusion of			Crystallization	
	Ta atoms to the sample surface	Si atoms to the sample surface	O atoms into the Cu/barrier interf.	Time	Products
$Ta_{73}Si_{27}$	1	n.o.	1	4 / 100	Ta_5Si_3 / Ta_2Si
$Ta_{62}Si_{20}N_{18}$	1	32	1	16 / 64	Ta_2N / Ta_5Si_3
$Ta_{56}Si_{19}N_{25}$	8	8	1	8	Ta_2N
$Ta_{41}Si_{20}N_{39}$	32	32	1	32	Ta_5N_6
$Ta_{30}Si_{18}N_{52}$	n.o.	n.o.	1	> 100	n.o.

4.2.2. Structure Changes for Cu-Capped Ta-Si-N Single Layers

To characterize the thermal stability of ternary Ta-Si-N diffusion barriers between copper and silicon oxide, annealing experiments were performed at $T = 600$ °C for $t = 1$ h ... 100 h and detailed investigations were carried out using complementary analytical techniques [237, 238]. It will be shown below that different processes contribute to the degradation of the Ta-Si-N barriers. Besides Ta diffusion to the sample surface and O diffusion into the Cu/barrier interface starting both before barrier crystallization, the formation of various Ta silicides and Ta nitrides is detected. As a consequence of crystalline phase formation, enhanced Ta, Si, and N interdiffusion as well as Cu diffusion into the substrate occur. The glancing angle XRD and GDOES results of this study are summarized in Table 5 and are discussed in more detail in the following.

Figure 15 (a) shows XRR curves for the $Cu/Ta_{73}Si_{27}/SiO_2/Si$ sample recorded after heat treatment at $T = 600$ °C. Significant changes of the barrier structure already occur during annealing for $t = 1$ h. In particular, an amplitude decrease of the barrier oscillations points to an increase of the Cu/barrier interface roughness. Additionally, the barrier layer thickness is reduced. In combination with the GDOES results (Table 5), both changes are primarily caused by disadvantageous Ta diffusion from the Ta-Si barrier along Cu grain boundaries to the sample surface. This diffusion process is comparable to that detected for a pure Ta barrier (cf. section 4.1.2). In the former case, however, the Ta diffusion is less intensive. Negative standard formation enthalpies for crystalline Ta silicides ($\Delta^f H^0_{Ta_2Si}$ = -126 kJmol^{-1}, $\Delta^f H^0_{TaSi_2}$ = -119 kJmol^{-1} [236]) are an indication for a stronger interaction between Ta and Si atoms than between Ta atoms among each other. Although Ta diffusion to the sample surface is also observed for a $Ta_{62}Si_{20}N_{18}$ layer during a 600 °C / 1h anneal (Table 5), its intensity is reduced compared to the $Ta_{73}Si_{27}$ barrier which can be explained with a smaller fraction of Ta-Ta bonds in the former case. TEM investigations indicate that Ta atoms diffused to the sample surface react with residual oxygen from the annealing ambience (surface layer B in Figure 16). Additional investigations show that the deposition of a SiN_x passivation layer onto the Cu film leads to reduced Ta diffusion into the SiN_x/Cu interface [239].

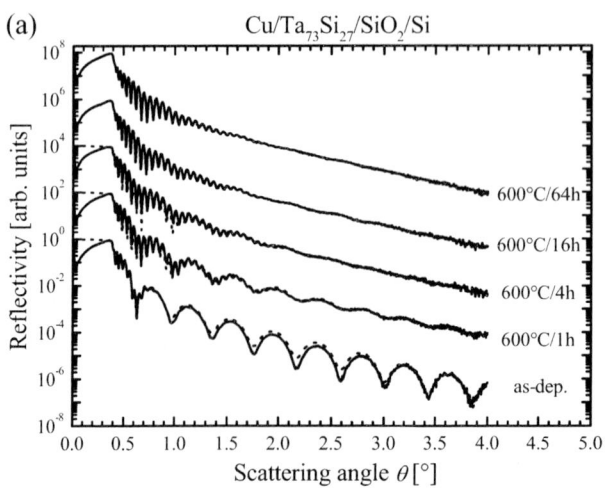

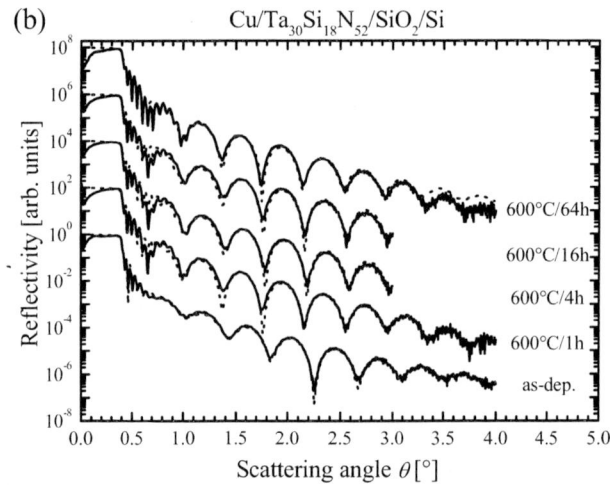

Figure 15 (a) reprinted from reference [237] with permission from Elsevier.

Figure 15. XRR curves for the Cu/Ta$_{73}$Si$_{27}$/SiO$_2$/Si sample (a) and the Cu/Ta$_{30}$Si$_{18}$N$_{52}$/SiO$_2$/Si sample (b) in the as-deposited state and after annealing at T = 600 °C for several times t (solid lines: experimental data, dotted lines: fit results).

Although there is no significant change of the Cu/barrier interface roughness, a slightly enhanced frequency of the barrier layer oscillations after annealing for t = 1 h at T = 600 °C (Figure 15 (b)) points to an increase of the Ta$_{30}$Si$_{18}$N$_{52}$ film thickness. According to a GDOES depth profile analysis, diffusion of O atoms from the residual gas of the annealing ambience along Cu grain boundaries into the Cu/Ta$_{30}$Si$_{18}$N$_{52}$ interface and the adjacent barrier region can account for the observed thickness increase. As shown by Shepherd et al. [240], exposure of an uncapped Ta-Si-N layer to air results in an oxidation of the near-surface region coupled with Ta and N depletion. Based on these findings, it can be assumed that O atoms diffused into the Cu/Ta-Si-N interface react with the upper part of the Ta-Si-N barrier. However, minor O diffusion into deeper barrier regions cannot be completely excluded. TEM bright field images taken after annealing at $T = 600$ °C for $t = 1$ h show homogeneous amorphous oxide layers for all Ta-Si-N barriers with a N content of $x_N \geq 25$ at.%. The oxide layers are, however, inhomogeneous for the samples with Ta$_{73}$Si$_{27}$ and Ta$_{62}$Si$_{20}$N$_{18}$ diffusion barriers, because in these two cases, O diffusion into the Cu/Ta-Si-N interface is superimposed by Ta diffusion to the sample surface.

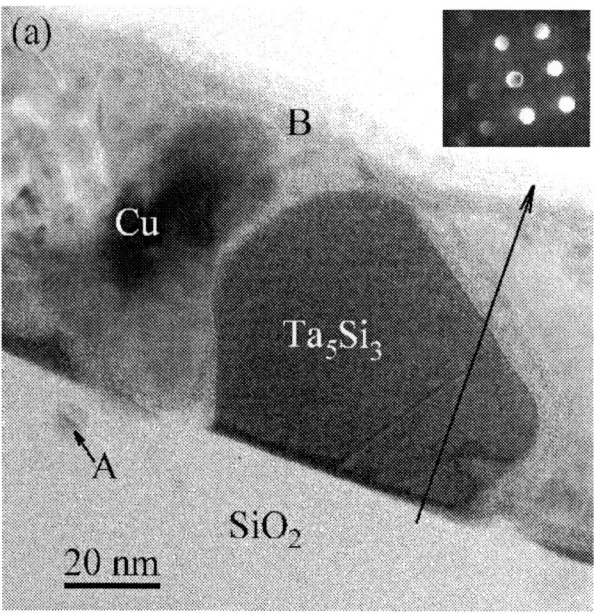

Figure 16 (Continued)

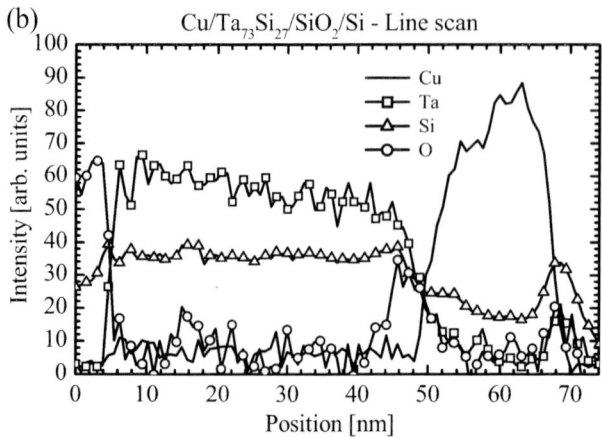

Figure 16. Cross-sectional TEM bright field image of the Cu/Ta$_{73}$Si$_{27}$/SiO$_2$/Si sample heat-treated at $T = 600$ °C for $t = 16$ h together with the convergent beam electron diffraction pattern for the Ta$_5$Si$_3$ crystallite (a). Part (b) shows the element distributions obtained by a combined EDXS and EELS line scan across the sample.

For Ta-Si-N barriers with $x_N \leq 39$ at.%, heat treatment for $1\ \text{h} < t < 100\ \text{h}$ at $T = 600$ °C leads to the formation of crystalline phases which requires an interdiffusion of various elements contained in the barrier. Concerning the annealing time necessary for crystallization (t_c), there is no unambiguous trend in the case of Ta-Si-N layers with $x_N \leq 25$ at.%. However, t_c increases for barriers with $x_N \geq 25$ at.% (Table 5). For the Cu/Ta$_{30}$Si$_{18}$N$_{52}$/SiO$_2$/Si sample, the amorphous microstructure remains stable even after annealing for $t = 100$ h. A comparison of the above conditions for crystalline phase formation with those for uncapped Ta-Si-N layers (cf. section 4.2.1) indicates that the presence of a Cu film accelerates the crystallization process, in particular for Ta-Si-N barriers with low N content. In the case of the Ta$_{56}$Si$_{19}$N$_{25}$/SiO$_2$/Si sample, for example, Ta$_2$N formation is observed after annealing at $T = 1000$ °C for $t = 1$ h, while the same crystalline phase is detected for the Cu/Ta$_{56}$Si$_{19}$N$_{25}$/SiO$_2$/Si sample after heat treatment at $T = 600$ °C for $t = 8$ h. In principal, there are three possibilities to explain accelerated barrier crystallization in the presence of a Cu film: (*i*) Copper diffuses into the amorphous Ta-Si-N layer and forms a quaternary compound with reduced crystallization temperature. In the recorded XRD patterns, there are, however, no indications for Cu compound formation. (*ii*) Particular atomic species leave the barrier region and diffuse into copper resulting in a lower crystallization temperature caused by slight changes of the Ta-Si-N composition. In this regard, the above-discussed Ta diffusion might contribute to an accelerated crystallization. (*iii*) Due to the presence of the Cu/barrier interface, heterogeneous nucleation is very likely. In this case, the nucleation energy is significantly reduced compared to homogeneous nucleation [241]. The role of minor impurities introduced into the interfaces during layer deposition remains, however, unclear. Regarding the crystallization products, Ta_5Si_3 and Ta_2Si are detected in the case of a binary $Ta_{73}Si_{27}$ layer. To motivate the growth of these two silicides, crystallization of an amorphous barrier is compared with cooling down a homogeneous melt [242]. Since the crystallization process is relatively slow at $T = 600$ °C, quasi-static cooling is assumed. Although sputter-deposited from a Ta_5Si_3 target, the chemical composition of the as-prepared Ta-Si layer is determined to $Ta_{73}Si_{27}$ which is in fairly good agreement to the stoichiometry of the Ta_3Si phase. However, the above-described Ta diffusion to the sample surface, which already starts before barrier crystallization, changes the x_{Si}/x_{Ta} ratio of the amorphous barrier. Thus, it is assumed that the Si content in the Ta-Si layer increases to a value between 33.3 at.% (Ta_2Si) and 37.5 at.% (Ta_5Si_3). According to the Ta-Si phase diagram [243], Ta_5Si_3 is the first crystalline phase forming during quasi-static cooling of a melt with such a composition. Further temperature decrease leads to Si depletion of the liquid and during a peritectic transformation the remaining melt reacts with Ta_5Si_3 to Ta_2Si. A precondition for all these considerations is, however, that the amorphous barrier neither reacts with the Cu metallization layer nor with dielectric SiO_2. The thermal stability with the substrate is ensured by the investigations presented in the previous subsection. In addition, almost constant absolute positions of the XRR curves in Figures 15 (a) and (b) point to unchanged barrier/SiO_2 interface roughnesses during annealing. Due to absent Bragg reflections of crystalline Cu compounds in the diffraction patterns, a reaction between copper and the amorphous barrier can be also excluded. Contrary to sole formation of Ta silicides (Ta_5Si_3 and Ta_2Si) for the Cu/$Ta_{73}Si_{27}$/SiO_2/Si sample, a Ta nitride is always the first crystallization product in the case of barriers containing nitrogen (Table 5). With increasing N content in the Ta-Si-N barrier, the crystalline Ta nitride gets N-richer ($Ta_{62}Si_{20}N_{18}$, $Ta_{56}Si_{19}N_{25}$: Ta_2N; $Ta_{41}Si_{20}N_{39}$: Ta_5N_6). For barriers with 0 at.% $< x_N <$ 25 at.%, Ta_5Si_3 is formed as second intermetallic phase. Figures 15 (a) and 16 show that annealing of an amorphous $Ta_{73}Si_{27}$ barrier for $t = 16$ h at $T = 600$ °C leads to its complete dissolution. Ta_5Si_3 crystallites are localized on silicon oxide and grow into the Cu film. Contrary to that, crystalline Ta nitrides are formed mainly within the original barrier region. For a $Ta_{62}Si_{20}N_{18}$ barrier, the Ta_5Si_3 crystallites do not grow on the SiO_2 substrate, but rather on the initially formed Ta_2N grains.

According to EELS line scans performed at cross-sectional Cu/$Ta_{62}Si_{20}N_{18}$/SiO_2/Si and Cu/$Ta_{56}Si_{19}N_{25}$/SiO_2/Si samples, silicon is not incorporated into the crystalline Ta nitrides. The main fraction leaves the original barrier region during the crystallization process. These Si atoms diffuse into the Cu layer where they are dissolved [244] or further to the sample surface where they are oxidized. Only a small Si fraction remains at is original position. TEM experiments show that there are local sites where silicon and oxygen are detected as the main components. Thus, it can be concluded that the crystalline Ta nitrides do not form a continuous layer.

Although diffusion of Ta atoms from the Ta-Si-N barrier to the sample surface is not detected for the Cu/$Ta_{56}Si_{19}N_{25}$/SiO_2/Si sample and the Cu/$Ta_{41}Si_{20}N_{39}$/SiO_2/Si sample at annealing for $t < 8$ h and $t < 32$ h, respectively, it occurs during barrier layer crystallization. Contrary to silicon, excess Ta atoms cannot be incorporated into the Cu film since the Ta solubility in copper is very low in the solid state [127]. A comparison of the Ta amounts diffused to the Cu surface for a particular annealing time indicates that the intensity of the Ta diffusion process decreases with increasing N content in the Ta-Si-N layer.

To prove Cu diffusion into the barrier and the adjacent SiO_2 substrate, glancing angle XRD investigations at Cu/barrier/SiO_2/Si samples are unsuitable, since Cu compound formation does not occur during annealing. GDOES depth profiles show at least an increased Cu signal in the barrier region after Ta-Si-N layer crystallization. As revealed by TEM bright field images, Cu grains can get into direct contact with SiO_2 during intensive heat treatment. Based on the fact that copper tends to agglomerate as a result of electron impact on the TEM specimen, EELS analyses confirm Cu diffusion into silicon oxide for the Cu/$Ta_{73}Si_{27}$/SiO_2/Si sample annealed at $T = 600$ °C for $t = 16$ h (dark-gray spot A in Figure 16 (a)) and the Cu/$Ta_{62}Si_{20}N_{18}$/SiO_2/Si sample annealed at $T = 600$ °C for $t = 100$ h. According to these findings, Cu atoms diffuse into the substrate at least after barrier crystallization. To clarify if Cu diffusion already occurs in the amorphous Ta-Si-N layers, trace-analytical techniques with an extremely low detection limit for copper, such as SIMS or AAS, can be applied. A change of the sample design, like a barrier deposition directly onto silicon, is another possibility to sensitively detect Cu diffusion through the barrier. Both approaches were applied in this study and will be discussed in the following two sections.

Chapter 5

TRACE-ANALYTICAL TECHNIQUES FOR A SENSITIVE PROOF OF CU DIFFUSION

Based on XRD, GDOES, and TEM analyses, microstructure changes occurring during thermal treatment of Ta-based diffusion barriers on SiO_2 were discussed in the previous section. However, regarding a systematic characterization of Cu diffusion through the barrier region into the dielectric substrate, the analytical techniques used so far are not sensitive enough. That is why trace-analytical methods with a very low detection limit for copper have to be applied. Using wet-chemical etching in combination with *atomic absorption spectrometry*, a quantitative Cu trace analysis is performed for each part of the $Cu/Ta_{56}Si_{19}N_{25}/SiO_2/Si$ sample after various annealing experiments (section 5.1). Based on additional *secondary ion mass spectroscopy* depth profile studies for the $Ta_{56}Si_{19}N_{25}/Cu/Ta_{56}Si_{19}N_{25}/SiO_2/Si$ sample (section 5.2), conclusions about the failure mechanism for a $Ta_{56}Si_{19}N_{25}$ barrier can be drawn.

5.1. Atomic Absorption Spectrometry (AAS)

Due to the low detection limits for copper in silicon ($2.5*10^{14}$ atomscm^3) and in SiO_2 ($5.0*10^{16}$ atoms/cm^3), AAS experiments were performed for the $Cu/Ta_{56}Si_{19}N_{25}/SiO_2/Si$ sample. As shown in Figure 17, annealing at $T = 400$ °C / $t = 100$ h and $T = 600$ °C / $t = 1$ h leads to an increase of the Cu content in the $Ta_{56}Si_{19}N_{25}$ barrier. Although explainable by commencing Cu diffusion, the enhanced Cu content may be also caused by AAS sample preparation. Residual O atoms which diffuse during annealing into the Cu/barrier interface might not only react with the barrier components (cf. section 4.2.2), but also with copper. However, such a mixed oxide layer is not soluble in ammonium persulfate. Consequently, it would be etched away together with the $Ta_{56}Si_{19}N_{25}$ film leading to an overestimation of the Cu content in the barrier. Significant Cu diffusion into silicon oxide and into silicon is detected after annealing at $T = 600\ °C$ for $t = 16\ h$ (Figure 17). In combination with the results of the previous section, it can be concluded that Cu diffusion into the SiO_2/Si substrate occurs at least after $Ta_{56}Si_{19}N_{25}$ layer crystallization.

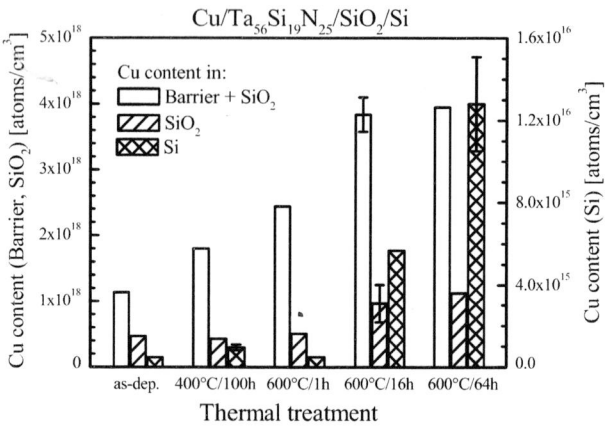

Figure 17. Cu concentrations determined by GF-AAS for the $Cu/Ta_{56}Si_{19}N_{25}/SiO_2/Si$ sample in the as-deposited state and after several heat treatments.

5.2. Secondary Ion Mass Spectroscopy (SIMS)

Besides time-consuming AAS experiments, which allow for a quantitative determination of Cu traces in laterally extended sample areas, SIMS investigations are also suitable for a sensitive proof of Cu diffusion through the barrier. With a detection limit of $2*10^{16}$ Cu atoms in 1 cm^3 silicon, SIMS depth profiles can be recorded for small sample areas in the order of 10^4 μm^2.

In this study, SIMS depth profile analysis was performed for a $Ta_{56}Si_{19}N_{25}/Cu/Ta_{56}Si_{19}N_{25}/SiO_2/Si$ layer stack. With this sample setup, wet-chemical etching of copper can be avoided and the measured Cu signal within the passivation layer is not significantly influenced by sputter artifacts. As shown in

Figure 18, curve (3), annealing of the $Ta_{56}Si_{19}N_{25}/Cu/Ta_{56}Si_{19}N_{25}/SiO_2/Si$ sample at $T = 650$ °C leads not only to a broadening of the Cu film signal, but also to an increase of the Cu intensity in the passivation and the barrier layer which can be explained with Cu trace diffusion. Additional TEM investigations [245] indicate that heat treatment at $T = 650$ °C leads to a degradation of the film integrity. In particular, there are sample positions where copper cannot be detected any longer. Instead, polycrystalline Ta compounds are observed there. Copper which was present in the as-deposited state at the considered sample positions diffuses into metallization regions adjacent to the Ta-based crystallites. As a consequence, the Cu film thickness increases locally which accounts for the Cu signal broadening detected by SIMS (Figure 18, curve (3)). Since grain boundaries are formed during $Ta_{56}Si_{19}N_{25}$ layer crystallization, Cu atoms can diffuse along them which results in a measurable Cu signal within both $Ta_{56}Si_{19}N_{25}$ layers (Figure 18, curve (3)).

Summarizing the trace-analytical studies, significant Cu diffusion could not be detected in an amorphous $Ta_{56}Si_{19}N_{25}$ layer. However, after barrier crystallization and, thus, grain boundary formation, Cu diffusion occurred.

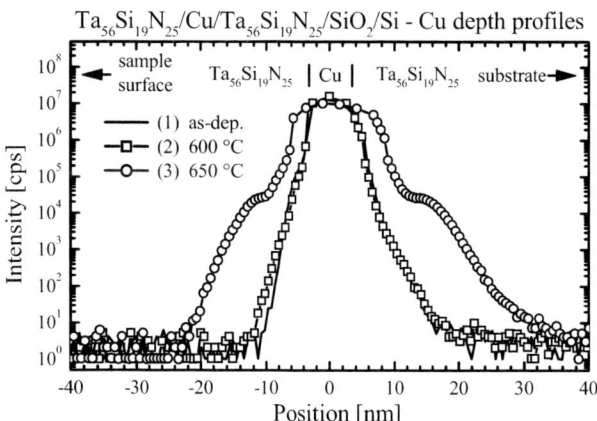

Reprinted from reference [245] with permission from Elsevier.

Figure 18. SIMS depth profiles of the Cu distribution for the $Ta_{56}Si_{19}N_{25}/Cu/Ta_{56}Si_{19}N_{25}/SiO_2/Si$ samples in the as-deposited state and heat-treated at several temperatures T for $t = 1$ h. The conversion of the sputter time into a length scale was done with the film thicknesses determined using XRR measurements in the as-deposited state. For comparison, the zero point of the length scale was shifted to the position of maximum Cu intensity.

6. THERMAL STABILITY OF TA-BASED DIFFUSION BARRIERS ON SILICON AND BARRIER FAILURE MECHANISMS

Besides the application of trace-analytical techniques, a sensitive proof of Cu diffusion into the substrate is also possible by XRD and partly by GDOES for samples with the diffusion barrier deposited directly onto silicon. For the extreme case of a missing barrier, Figure 2 indicated that Cu_3Si formation is detected at temperatures as low as $T \approx 190\,°C$. Besides this low reaction temperature, the good detectability of the copper silicide by XRD is an additional advantage for proving Cu diffusion through the barrier into silicon. However, to better understand the microstructure changes for the Cu/barrier/Si samples during annealing, the characterization of the barrier-Si reaction behavior is necessary. For that reason, barrier/Si samples are investigated prior to the analysis of the Cu-coated layer stacks.

6.1. Ta-TaN Layer Stacks

6.1.1. Reaction Behavior with Silicon

In the case of a _Ta layer_ deposited directly onto silicon, the same microstructure changes occur during low-temperature annealing as for the $Ta/SiO_2/Si$ sample (cf. section 4.1.1). In particular, an increase of the intrinsic compressive stress, diffusion of C atoms into tantalum, and the β'-Ta nucleation are observed. However, at $T = 550\,°C$, an amorphous Ta-Si intermixing layer is formed between tantalum and silicon. The growth of $TaSi_2$ crystallites at the interface to the substrate is detected for the first time at $T = 620\,°C$ [220]. These findings are in good agreement with experimental results by Cheng et al. [130] and Nguyen Tan et al. [246] as well as theoretical considerations by Walser et al. [247] who published a rule that the first compound nucleated in a planar silicon – transition metal binary couple is the most stable congruently melting compound adjacent to the lowest-temperature eutectic in the bulk equilibrium phase diagram. Thus, the reaction of a Ta layer on a Si substrate should result in the formation of hexagonal $TaSi_2$ as the first crystalline Ta silicide. Instead of $TaSi_2$, Noya et al. [248, 249] detected Ta_5Si_3. Although for infinite reservoirs of Ta and Si atoms, equilibrium thermodynamics would favor the formation of Ta_5Si_3 over that of $TaSi_2$ (Gibbs energy of formation per mol educts at $T = 427\,°C$: $\Delta^f G_{Ta_5Si_3} = -43.2\ \mathrm{kJmol^{-1}}$, $\Delta^f G_{TaSi_2} = -38.9\ \mathrm{kJmol^{-1}}$ [236]), the first crystalline Ta silicide phase seems to be selected by kinetic processes [250]. According to investigations by Baglin et al. [251], it can be assumed that silicon is the diffusant in the Ta-Si solid state reaction. At its beginning, there is no nucleation of a crystalline Ta silicide. Rather, an amorphous Ta-Si intermixing layer is formed due to a stronger decrease of the system's free enthalpy per unit time. The thickness of the intermixing layer increases linearly with time [130]. This means that Si diffusion through the amorphous Ta-Si film is faster than the delivery of Si atoms at the substrate interface. If, however, the intermixing layer has reached a certain thickness, its growth rate will slow down because the Si concentration gradient and, thus, the Si diffusion rate in the Ta-Si film decrease. Eventually, the formation of a crystalline Ta silicide at the Ta-Si/Si interface gets energetically more favored. Up to $T = 900$ °C, $TaSi_2$ is the only crystalline Ta silicide observed in the diffraction diagram (Table 6). The fact that there is no additional Ta-Si compound formation is in good accordance with theoretical considerations by Tsaur et al. [252].

Table 6. Phase formation behavior for the Ta/Si and the TaN/Si system during annealing at temperatures between 500 °C $\leq T \leq$ 1100 °C for $t = 1$ h

	Ta/Si	TaN/Si
as-depos.	β-Ta	TaN, amorphous
500°C / 1h	β-Ta	TaN, amorphous
550°C / 1h	β-Ta, β'-Ta	- - -
600°C / 1h	β-Ta, β'-Ta	TaN, amorphous
650°C / 1h	β-Ta, β'-Ta, $TaSi_2$, Ta_2C	-
700°C / 1h	β'-Ta, $TaSi_2$, Ta_2C	TaN, amorphous
800°C / 1h	β'-Ta, $TaSi_2$, Ta_2C	TaN, amorphous
900°C / 1h	$TaSi_2$, Ta_2C	TaN, amorphous
1000°C / 1h	-	Ta_4N_5
1050°C / 1h	-	$TaSi_2$
1100°C / 1h	-	$TaSi_2$

Samples belonging to boxes marked with an "-" were not investigated.

For a *TaN layer* deposited onto silicon, heat treatment up to $T = 1000$ °C leads to microstructure changes comparable to those for silicon oxide as substrate material. In particular, fcc TaN crystallite growth is observed up to $T = 900$ °C and tetragonal Ta_4N_5 formation occurs at $T = 1000$ °C (Tables 2 and 6). Annealing at $T = 1050$ °C results in a reaction between Ta_4N_5 and silicon to $TaSi_2$. A comparison with the $TaN/SiO_2/Si$ sample suggests that this reaction is associated with the transformation of tetragonal Ta_4N_5 into hexagonal Ta_5N_6. At this phase transformation, the Ta coordination is changed. Although still bonded to six N atoms, the latter ones do not necessarily form an octahedron for Ta_5N_6. Consequently, Ta-N bonds are broken during the transformation between both Ta nitrides wherewith Si atoms get the chance to react with tantalum.

6.1.2. Structure Changes for Cu-Capped Ta-TaN Layer Stacks

Heat treatment of a *Ta diffusion barrier* between copper and silicon at $T = 550$ °C leads to a significant increase of the amorphous phase content. This effect, which was not observed for the $Cu/Ta/SiO_2/Si$ sample, might be caused by the formation of an amorphous Ta-Si intermixing layer between tantalum and silicon (cf. section 6.1.1). Additionally, first signs of Cu_3Si formation point to Cu diffusion through the Ta barrier (Table 7). GDOES measurements are in accordance with these XRD findings. Besides a Ta surface signal which is the result of Ta diffusion through copper (Figure 19 (b), curve (3)), there is an increase of the Cu intensity in the barrier and the substrate region (Figure 19 (a), curve (3)) after annealing at $T = 550$ °C. Gray punctual spots, which are statistically distributed across the Cu surface and are visible with naked eyes, give additional evidence for a local diffusion of Cu atoms through the Ta film into the Si substrate resulting in an uncovering of the barrier layer.

The findings concerning the commencing Cu diffusion through a 10 nm thick Ta barrier at $T = 550$ °C are in good agreement with results by Hecker et al. [78] and Laurila et al. [62]. Due to the nanocrystalline microstructure of tetragonal β-Ta with grain sizes in the order of the film thickness (cf. section 3.1), there are grain boundaries which can act as fast paths for Cu diffusion. Based on the corresponding diffusion parameters ($D'_{0,Cu/Ta} = 9.0*10^{-4}$ cm^2s^{-1} and $Q'_{Cu/Ta} = 2.3$ eV [165]), the mean diffusion distance $l = (Dt)^{1/2}$ is calculated to $l_{Cu/Ta} \approx 2$ nm at $T = 550$ °C and $t = 1$ h. This is slightly smaller than the as-deposited Ta film thickness but not in contradiction to the results of this study, because local thinning of the Ta barrier occurs already at $T = 500$ °C due to Ta diffusion to the sample surface.

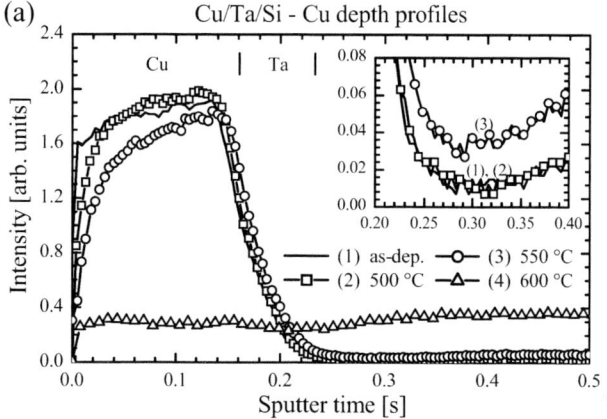

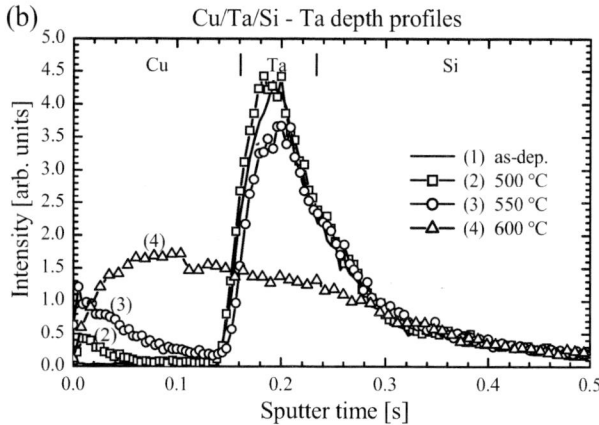

Figure 19. GDOES depth profiles of the Cu distribution (a) and of the Ta distribution (b) for the Cu/Ta/Si sample in the as-deposited state and after annealing at several temperatures T for $t = 1$ h.

According to the XRD analysis, Cu_3Si formation is detected prior to $TaSi_2$ nucleation which occurs at $T = 575$ °C. This Ta silicide formation temperature is at least 45 K lower compared to the sample without Cu cap layer. Thus, it is concluded that Cu diffusion into silicon promotes $TaSi_2$ grain growth. As discussed in the previous subsection, annealing at $T \geq 500$ °C initially leads to the formation of an amorphous film at the Ta/Si interface. The growth rate of this Ta-Si intermixing layer is limited by the delivery of free Si atoms. Nucleation of crystalline $TaSi_2$ does not occur until $T = 620$ °C. In contrast to that, Cu_3Si formation at the Cu/Si interface is determined by Cu diffusion into the Si substrate. There, Cu atoms occupy interstitial lattice sites which leads to an increased number of nearest neighbors for the adjacent Si atoms and, thus, to a charge transfer and a decreased Si-Si bond strength. As a consequence, Si atoms leave their regular lattice sites, diffuse as interstitials to the Cu/Si interface, and react with copper to Cu$_3$Si [253]. Since for the Cu/Ta/Si sample, Cu silicide nucleation occurs prior to Ta silicide formation, sufficient free Si atoms are present at the substrate interface to accelerate the growth of the amorphous Ta-Si intermixing layer which finally results in crystalline TaSi$_2$ nucleation at a lower temperature.

Contrary to a Ta layer with β-Ta crystallite sizes in the order of the film thickness, the microstructure of a *TaN diffusion barrier* is characterized by small fcc TaN grains ($d_{TaN} \approx 3$ nm, section 3.1) in an amorphous matrix. Thus, there are no grain boundaries running directly from the Cu/TaN interface to the TaN/substrate interface. Heat treatment at elevated temperatures leads, however, to TaN grain growth. Although the mean crystallite size remains smaller than the TaN film thickness, several TaN grain boundaries connecting the Cu metallization layer with the Si substrate and enabling fast Cu diffusion seem to exist. Based on the grain boundary diffusion parameters $D'_{0,Cu/TaN} = 2.8*10^{-10}$ cm^{2}s^{-1} and $Q'_{Cu/TaN} = 1.3$ eV published by Oku [181], a mean diffusion distance of $l_{Cu/TaN} \approx 9$ nm is estimated for annealing at $T = 800$ °C and $t = 1$ h. This result is in good agreement with the experimental findings of this study. Commencing Cu$_3$Si formation and, thus, significant Cu diffusion through a 10 nm thick TaN layer into silicon occur during an equivalent thermal treatment (Figure 20). As in the case of the Cu/Ta/Si sample, the Cu diffusion process leads to the generation of sufficient free Si atoms which can diffuse along grain boundaries into the TaN layer and finally react with the barrier to form TaSi$_2$. Compared to the sample without Cu cap layer, this reaction temperature is thus lowered by $\Delta T \approx 200$ K (Tables 6 and 7).

A *Ta/TaN/Ta diffusion barrier* deposited between copper and silicon shows the same low-temperature annealing behavior as the Cu/Ta/TaN/Ta/SiO$_2$/Si sample (cf. section 4.1.2). Nitrogen diffusion from the TaN layer into both adjacent Ta films starts at $T = 300$ °C and leads to hexagonal Ta$_2$N formation at $T = 500$ °C (Table 7). For the Cu/Ta/TaN/Ta/Si sample, nucleation of hexagonal TaSi$_2$ occurs at $T = 650$ °C. This reaction temperature is comparable to the value determined for the Ta/Si sample. Thus, TaSi$_2$ nucleation in the case of the Cu-coated sample is not catalyzed by Cu diffusion. However, since Cu$_3$Si formation does not occur until $T = 700$ °C, grain boundaries due to TaSi$_2$ formation may slightly accelerate Cu diffusion into silicon.

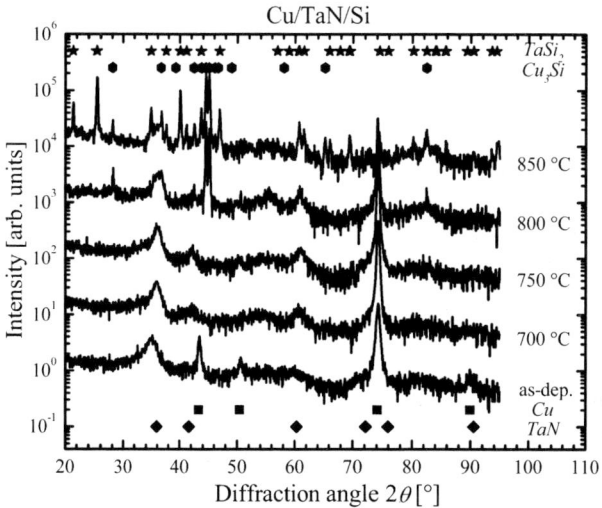

Figure 20. Glancing angle XRD diagrams of the Cu/TaN/Si sample in the as-deposited state and after annealing at several temperatures T for $t = 1$ h.

Table 7. Crystalline phase formation behavior for different Cu/barrier/Si systems during annealing at temperatures between 300 °C ≤ T ≤ 850 °C for $t = 1$ h

	Cu/Ta/Si	Cu/TaN/Si	Cu/Ta/TaN/Ta/Si
as-depos.	Cu, β-Ta	Cu, TaN	Cu, α-Ta, TaN, β-Ta
300°C / 1h	-	-	Cu, α-Ta, TaN, β-Ta
400°C / 1h	-	-	Cu, α-Ta, TaN, β-Ta
500°C / 1h	Cu, β-Ta	Cu, TaN	Cu, α-Ta, TaN, β-Ta, Ta$_2$N
550°C / 1h	Cu, Cu$_3$Si, β-Ta	Cu, TaN	-
600°C / 1h	Cu$_3$Si, β-Ta, TaSi$_2$	Cu, TaN	Cu, Ta$_2$N
650°C / 1h	Cu$_3$Si, TaSi$_2$	Cu, TaN	Cu, Ta$_2$N, TaSi$_2$
700°C / 1h	Cu$_3$Si, TaSi$_2$	Cu, TaN	Cu$_3$Si, Ta$_2$N, TaSi$_2$
750°C / 1h	-	Cu, TaN	Cu$_3$Si, Ta$_2$N, TaSi$_2$
800°C / 1h	-	Cu, Cu$_3$Si, TaN	Cu$_3$Si, TaSi$_2$
850°C / 1h	-	Cu$_3$Si, TaN, TaSi$_2$	-

Samples belonging to boxes marked with an "-" were not investigated.

Based on the critical temperatures for Cu_3Si formation determined by glancing angle X-ray diffraction, the following relation (12) holds for the thermal stabilities ($S_{barrier}$) of the corresponding diffusion barriers:

$$S_{Ta} < S_{Ta/TaN/Ta} < S_{TaN}. \tag{12}$$

In particular, relation (12) implies that a 30 nm thick Ta/TaN/Ta barrier is characterized by a lower thermal stability than a 10 nm thick TaN film which is part of the trilayer. This result can be explained with the N redistribution in the Ta-TaN layer stack and the subsequent Ta_2N formation which lead to a degradation of the thermally stable TaN layer. Continuative experiments should clarify, whether TaN layers with $x_N/x_{Ta} \leq 1$ can prevent N diffusion into the adjacent Ta films.

6.2. Ta-Si-N Single Layers

6.2.1. Reaction Behavior with Silicon

Although there are no visible changes in the XRD pattern, annealing at $T = 600\ °C$ leads to Si diffusion from the substrate into the $\underline{Ta_{73}Si_{27}\ layer}$. Nucleation of tetragonal Ta_5Si_3 is observed at $T = 650\ °C$ (Table 8). According to XRR and GDOES investigations, this Ta silicide grows within a continuous layer. Since tetragonal Ta_5Si_3 is also found as crystallization product for the $Cu/Ta_{73}Si_{27}/SiO_2/Si$ sample (Table 5), it is assumed that Si diffusion from the substrate increases the x_{Si}/x_{Ta} ratio of the amorphous $Ta_{73}Si_{27}$ layer which leads to its accelerated crystallization. The initially formed Ta_5Si_3 crystallites are not stable. At elevated temperatures, they react with silicon to hexagonal $TaSi_2$ (Table 8). According to the XRD results, the transformation of the amorphous into a crystalline microstructure occurs for the Ta-Si-N films during annealing at $T > 900\ °C$. In the case of a $\underline{Ta_{56}Si_{19}N_{25}\ layer}$, hexagonal Ta_2N and Ta_5Si_3 are nucleated at $T = 950\ °C$. The crystallization seems to occur with a small amount of silicon from the substrate, since Ta silicide formation could not be detected for the $Ta_{56}Si_{19}N_{25}/SiO_2/Si$ sample (Table 4). For a $\underline{Ta_{30}Si_{18}N_{52}\ layer}$, fcc TaN formation is observed at $T = 1000\ °C$ and a significant reaction with silicon occurs at $T = 1050\ °C$ (Table 8).

Table 8. Phase formation behavior for Ta-Si-N/Si systems during annealing at temperatures between 600 °C ≤ T ≤ 1100 °C for t = 1 h

	$Ta_{73}Si_{27}$/Si	$Ta_{56}Si_{19}N_{25}$/Si	$Ta_{30}Si_{18}N_{52}$/Si
as-depos.	amorphous	amorphous	amorphous
600°C / 1h	amorphous	amorphous	amorphous
650°C / 1h	amorphous, Ta_5Si_3	-	-
700°C / 1h	Ta_5Si_3	-	-
750°C / 1h	Ta_5Si_3, $TaSi_2$	-	-
800°C / 1h	$TaSi_2$	-	-
900°C / 1h	$TaSi_2$	amorphous	amorphous
950°C / 1h	-	amorphous, Ta_5Si_3, Ta_2N	amorphous
1000°C / 1h	-	Ta_5Si_3, Ta_2N, $TaSi_2$	amorphous, TaN
1050°C / 1h	-	$TaSi_2$	$TaSi_2$
1100°C / 1h	-	$TaSi_2$	$TaSi_2$

Samples belonging to boxes marked with an "-" were not investigated.

According to the above results, the amorphous Ta-Si-N microstructure is degraded by crystallization. As in the case of the $Ta_{73}Si_{27}$/Si sample, this process can be significantly accelerated by diffusion of Si atoms from the substrate into the Ta-based layer. With increasing N content in the Ta-Si-N layer, the impact of the Si substrate on the crystallization process is reduced. For a $Ta_{30}Si_{18}N_{52}$ layer, TaN formation is observed at the same temperature for a Si and a SiO_2 substrate. More intensive heat treatment leads to $TaSi_2$ nucleation. According to the XRD results, the propensity for a reaction with silicon decreases with increasing N content in the Ta-Si-N layer. However, for a film with a particular composition, the reaction with silicon is always detected at a lower temperature than the reaction with silicon oxide (cf. Tables 4 and 8).

6.2.2. Structure Changes for Cu-Capped Ta-Si-N Single Layers

Heat treatment of a *$Ta_{73}Si_{27}$ diffusion barrier* at T = 575 °C only leads to changes of the Cu texture, while crystalline Cu_3Si formation and, consequently, Cu diffusion into silicon occur at T = 590 °C. Additional $TaSi_2$ nucleation is detected at T = 600 °C (Figure 21 (a)). Tetragonal Ta_5Si_3, which is formed during annealing of the $Ta_{73}Si_{27}$/Si sample, is not observed. A *$Ta_{56}Si_{19}N_{25}$ diffusion barrier* deposited between copper and silicon shows an annealing behavior which is equivalent to that of the Cu/$Ta_{73}Si_{27}$/Si sample. Simultaneous formation of

Cu$_3$Si and TaSi$_2$ is detected at $T = 630$ °C. In the case of a *Ta$_{30}$Si$_{18}$N$_{52}$ diffusion barrier*, fist signs of Cu$_3$Si nucleation are visible after annealing at $T = 700$ °C. In contrast to the two other barriers, the critical temperature for TaSi$_2$ formation ($T = 900$ °C) is, however, significantly higher than that for Cu$_3$Si nucleation (Figure 21 (b)). It should be emphasized that apart from Cu$_3$Si and TaSi$_2$, no other crystalline phases are formed during heat treatment of the Cu/Ta-Si-N/Si samples.

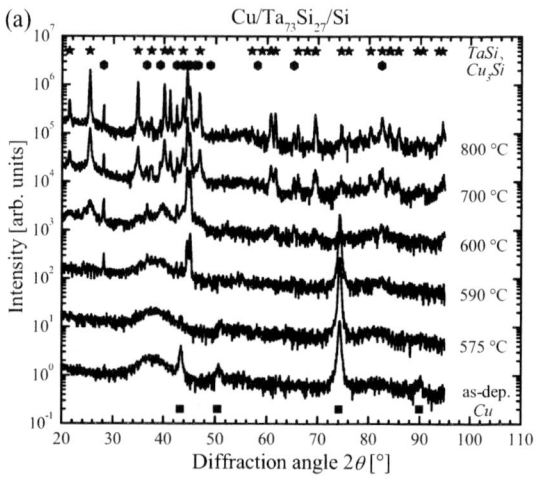

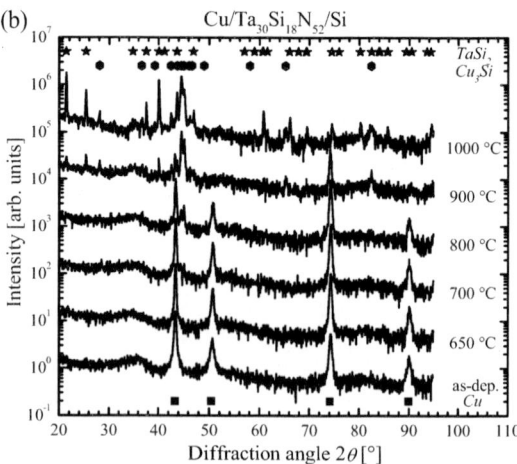

Figure 21. Glancing angle XRD diagrams of the Cu/Ta$_{73}$Si$_{27}$/Si sample (a) and the Cu/Ta$_{30}$Si$_{18}$N$_{52}$/Si sample (b) in the as-deposited state and after annealing at several temperatures T for $t = 1$ h.

Since Cu silicide Bragg reflections are not observed in the diffraction patterns for the Cu/Ta-Si-N/SiO$_2$/Si samples (cf. section 4.2.2), Cu$_3$Si formation in the case of the Ta-Si-N layers deposited directly onto silicon is the result of a reaction between copper and Si atoms from the substrate. If the latter ones would diffuse through the Ta-Si-N film into copper, they should be initially dissolved in the metallization layer [244]. Since this process does not occur, detection of Cu$_3$Si formation using glancing angle XRD is a suitable method to prove Cu diffusion into silicon. For Ta–Si–N films deposited between copper and silicon, the critical temperatures for Cu$_3$Si formation are always lower than the crystallization temperatures for the corresponding samples without Cu cap layer (Table 8). Thus, the Cu$_3$Si formation temperatures can be used to compare the thermal stabilities $S_{barrier}$ of the Ta-Si-N layers. Including the results for 10 nm thick Ta and TaN barriers (cf. section 6.1.2), the following relation (13) is obtained:

$$S_{Ta} < S_{Ta_{73}Si_{27}} < S_{Ta_{56}Si_{19}N_{25}} < S_{Ta_{30}Si_{18}N_{52}} < S_{TaN}. \tag{13}$$

In particular, the inequality implies that an amorphous Ta-Si diffusion barrier is characterized by a higher thermal stability than a polycrystalline Ta layer. Further improvement can be achieved by depositing Ta-Si-N films. However, for comparable N contents, a Ta$_{30}$Si$_{18}$N$_{52}$ layer is less stable against Cu diffusion than a TaN barrier.

Besides the determination of the thermal stabilities, the characterization of the failure mechanisms for the amorphous Ta-Si-N barriers is of special interest. In the case of a Ta$_{73}$Si$_{27}$ film deposited between copper and silicon, Cu$_3$Si formation and, thus, Cu diffusion are detected for the first time under comparable annealing conditions as the commencing crystallization of the same barrier prepared onto silicon oxide. Therefore, it can be assumed that a Ta$_{73}$Si$_{27}$ diffusion barrier fails by crystallization and associated defect formation. The slightly lower Cu$_3$Si nucleation temperature for the Cu/Ta$_{73}$Si$_{27}$/Si sample (T = 590 °C) compared to the crystallization temperature for the Cu/Ta$_{73}$Si$_{27}$/SiO$_2$/Si sample (T = 600 °C) might be explained by simultaneous Ta diffusion from the barrier to the sample surface and Si diffusion from the Si substrate into the barrier layer which in combination lead to accelerated barrier crystallization. After diffusing along grain boundaries, Cu atoms react with silicon to form Cu$_3$Si. During this process, free Si atoms are generated. They diffuse into the crystallized barrier and accelerate the formation of hexagonal TaSi$_2$.

For a Ta$_{56}$Si$_{19}$N$_{25}$ barrier, Cu$_3$Si and TaSi$_2$ formation are detected during the same annealing. In particular, the TaSi$_2$ nucleation temperature is thus reduced by at least $\Delta T = 370$ K in the presence of copper. Furthermore, Cu_3Si formation occurs at the same thermal treatment as crystallization of the $Ta_{56}Si_{19}N_{25}$ layer. Consequently, Cu atoms seem to diffuse along grain boundaries. These considerations are confirmed by the SIMS and AAS analyses presented in section 5.

In the case of a $Ta_{30}Si_{18}N_{52}$ barrier, Cu_3Si formation and, consequently, Cu diffusion are observed at significantly lower thermal stresses than barrier crystallization, which occurs for the $Cu/Ta_{30}Si_{18}N_{52}/SiO_2/Si$ sample at $T = 900$ °C. An equivalent annealing temperature is necessary for $TaSi_2$ nucleation in the case of the $Cu/Ta_{30}Si_{18}N_{52}/Si$ sample (Figure 21 (b)). If there were already massive barrier defects at commencing Cu_3Si formation, Si atoms should be able to diffuse along them and to react with the barrier, as in the case of a $Ta_{56}Si_{19}N_{25}$ layer. Since this behavior is not observed, Cu atoms seem to diffuse through an amorphous $Ta_{30}Si_{18}N_{52}$ barrier. These considerations are in good accordance with experimental results by Reid et al. [254] for a $Ta_{36}Si_{14}N_{50}$ barrier.

As in the case of Ta and TaN layers, $Ta_{73}Si_{27}$ and $Ta_{30}Si_{18}N_{52}$ films can be combined to layer stacks. Although the critical temperature for N diffusion from the Ta-Si-N film into the adjacent Ta-Si layers is increased to $T = 500$ °C, there is no stability improvement for the graded $Ta_{73}Si_{27}$-$Ta_{30}Si_{18}N_{52}$ barriers compared to a $Ta_{30}Si_{18}N_{52}$ single layer [255].

Chapter 6

CONCLUSION

Using magnetron sputtering, thin Ta and TaN single layers, Ta-TaN layer stacks, as well as Ta-Si-N single layers were deposited with and without Cu metallization onto blanket and thermally oxidized Si wafers. To characterize the thermal stability and the failure mechanisms of these Ta-based diffusion barriers, complementary analytical techniques, including X-ray scattering, spectroscopic methods, and microscopic analyses, were applied prior and after thermal treatment.

Due to Ta_2O_5 formation temperatures of $T \geq 900$ °C, all Ta-based layers meet the necessary precondition of sufficient thermal stability against a reaction with silicon oxide. Depending on the chemical composition and, consequently, on the as-deposited microstructure, the various diffusion barriers show a different annealing behavior when deposited between copper and silicon oxide. In the case of Ta-TaN layer stacks, nitrogen redistribution within the barriers starts at $T = 300$ °C and finally leads to Ta_2N formation at $T = 500$ °C. For Ta-Si-N films, the presence of a Cu metallization layer results in an accelerated crystallization of the barriers, whereby their compositions not only determine the annealing time to start the crystallization, but also the crystallization products. As a consequence of Ta-Si-N barrier crystallization, the original layer integrity is completely destroyed.

The sensitive proof of Cu diffusion through the barrier films was carried out based on two different approaches. Firstly, trace-analytical techniques with extremely low detection limits for copper, such as atomic absorption spectrometry, were applied at $Cu/barrier/SiO_2/Si$ structures. Secondly, samples with alternate film stacks were analyzed. In particular, barriers deposited directly between copper and silicon allowed for a sensitive proof of Cu diffusion via the detection of Cu_3Si formation by X-ray diffraction. Based on the critical temperatures for Cu silicide nucleation, the thermal stabilities of the diffusion barriers ($S_{barrier}$) are characterized as follows:

$$S_{Ta} < S_{Ta/TaN/Ta} < S_{TaN}$$

and

$$S_{Ta} < S_{Ta_{73}Si_{27}} < S_{Ta_{56}Si_{19}N_{25}} < S_{Ta_{30}Si_{18}N_{52}} < S_{TaN} \quad (14)$$

Considering both the thermal stability as well as the resistivity of the Ta-based layers, a $Ta_{56}Si_{19}N_{25}$ film seems to be the most suitable diffusion barrier for an application in Cu metallization systems.

The microstructure investigations of this study also allow for the characterization of the barrier failure mechanisms. For a TaN single layer, Ta nitride crystallite growth results in grain boundary elongation which finally enables Cu diffusion through the barrier. In the case of the investigated Ta-TaN layer stacks, N redistribution and subsequent Ta_2N formation destroy the stable nanocrystalline TaN microstructure already at low annealing temperatures. Further experiments should prove if a changed composition of the Ta nitride layer can prevent N diffusion into the adjacent Ta films. For Ta-Si-N films with a N content of $x_N \leq 25$ at.%, Cu diffusion into the substrates is only detected after barrier crystallization. In the case of N-rich Ta-Si-N films, however, Cu diffusion occurs before massive barrier crystallization starts.

Due to the continuous scaling down of all feature sizes, including the Cu interconnect dimensions, the thickness of the diffusion barrier layers has to be further reduced in the future. Since PVD-based techniques may not provide sufficient step coverage, advanced deposition methods have to be applied. Atomic layer deposition is considered to be one of the most promising techniques. The deposition of highly conformal and reliable diffusion barriers into narrow via and trench structures based on porous low-k and ultra low-k materials remains, however, a big challenge. Even for established materials, the diffusion barrier properties, like the thermal stability, can significantly change in combination with new deposition techniques and new interlayer dielectrics. Thus, sophisticated analytical techniques will be necessary in the future to carry out detailed microstructure investigations for a better understanding of the functional properties of ultrathin diffusion barriers.

ACKNOWLEDGMENTS

The results of this study were obtained within a joint BMBF project between the Materials Analysis Laboratory at AMD Saxony in Dresden, the Leibniz Institute for Solid State and Materials Research Dresden (IFW), the Semiconductor and Microsystems Technology Laboratory (IHM) at the Dresden University of Technology, and the Center for Microtechnologies (ZfM) at the Chemnitz University of Technology in Germany. For experimental support as well as stimulating discussions, the author would like to thank Ehrenfried Zschech, Michael Hecker, Hans-Jürgen Engelman, Daniel Gehre (all with AMD Saxony), Klaus Wetzig, Norbert Mattern, Volker Hoffmann, Rainer Reiche, Jörg Acker, Ingrid Wetzig, Andrea Voss, Siegfried Neumann (all with IFW), Johann W. Bartha, Christian Wenzel, Henning Heuer (all with IHM), as well as Thomas Geßner, Stefan Schulz, and Ramona Ecke (all with ZfM). Financial support from the BMBF (contr. No. 03N1067D) is gratefully acknowledged.

REFERENCES

[1] Greenwood, N. N.; Earnshaw, A. *Chemistry of the elements;* 2nd edition; Butterworth-Heinemann: Oxford, 1997; p 1176.
[2] G. A. Sai-Halasz, *Directions in future high end processors*, International Conference on Computer Design, VLSI in Computers and Processors, Cambridge, MA, U.S.A., October 11-14, 1992, IEEE Computer Soc. Proc. (1992) 230-233.
[3] Lide, D. R. *CRC Handbook of Chemistry and Physics;* 86th edition 2005-2006; CRC Press, Taylor & Francis Group: Boca Raton, FL, 2005; p 12-196.
[4] Hu, C.-K.; Harper, J. M. E. *Mater. Chem. Phys.* 1998, *52,* 5-16.
[5] Ogawa, E. T.; Lee, K.-D.; Blaschke, V. A.; Ho, P. S. *IEEE Trans. Reliab.* 2002, *51,* 403-419.
[6] Parkes, G. D. *Mellor's Modern Inorganic Chemistry;* Longmans, Green and Co LTD: London, 1967; p 650.
[7] Istratov, A. A.; Weber, E. R. *J. Electrochem. Soc.* 2002, *149,* G21-G30.
[8] Steinhögl, W.; Schindler, G.; Steinlesberger, G.; Engelhardt, M. *Phys. Rev. B* 2002, *66,* 075414/1-075414/4.
[9] Steinlesberger, G.; Engelhardt, M.; Schindler, G.; Steinhögl, W.; von Glasow, A.; Mosig, K.; Bertagnolli, E. *Microelectron. Eng.* 2002, *64,* 409-416.
[10] M. Engelhardt, G. Schindler, M. Traving, A. Stich, Z. Gabric, W. Pamler, W. Hönlein, *Scaling of Metal Interconnects: Challenges to Functionality and Reliability* in: E. Zschech, K. Maex, P. S. Ho, H. Kawasaki, T. Nakamura (Eds.), Eighth International Workshop on Stress-Induced Phenomena in Metallization, Dresden, Germany, September 12-14, 2005, AIP Conference Proceedings 817 (2006) 3-12.
[11] Wang, S.-Q. *MRS Bulletin* 1994, *19,* 30-40.

[12] Kaloyeros, A. E.; Eisenbraun, E. *Ann. Rev. Mater. Sci.* 2000, *30*, 363-385.
[13] Zschech, E. In *Metal Based Thin Films for Electronics;* Wetzig, K.; Schneider C. M.; Eds.; Wiley-VCH GmbH & Co. KGaA: Weinheim, 2003; pp 222-235.
[14] *The International Technology Roadmap for Semiconductors (ITRS),* Semiconductor Industry Association (SIA), 2005.
[15] Philibert, J. *Atom movements, Diffusion and mass transport in solids;* Les Éditions de Physique, 1991.
[16] Crank, J. *The mathematics of diffusion;* Oxford University Press: London, 1964; p 11.
[17] Harrison, L. G. *Trans. Faraday Soc.* 1961, *57*, 1191-1199.
[18] Beke, D. L. In Landolt-Börnstein: Numerical Data and Functional Relationships in Science and Technology - New Series, Group 3: Condensed Matter; Springer-Verlag: Heidelberg, 1999; Vol. 33: Diffusion in Semiconductors and Non-Metallic Solids, Subvol. B: Diffusion in Non-Metallic Solids, Part 1, pp 1-23.
[19] Kaur, I.; Gust, W. *Fundamentals of grain and interphase boundary diffusion;* Ziegler Press: Stuttgart, 1989; p 79.
[20] Fisher, J. C. *J. Appl. Phys.* 1951, *22*, 74-77.
[21] Whipple, R. T. P. *Phil. Mag.* 1954, *45*, 1225-1236.
[22] Suzuoka, T. *Trans. Jpn. Inst. Met.* 1961, *2*, 25-33.
[23] Suzuoka, T. *J. Phys. Soc. Jpn.* 1964, *19*, 839-851.
[24] Le Claire, A. D.; Rabinovitch, A. *J. Phys. C: Solid State Phys.* 1981, *14*, 3863-3879.
[25] Gjostein, N. A. In *Diffusion*; American Society for Metals: Metals Park, OH, 1973; p 241.
[26] Istratov, A. A.; Flink, C.; Hieslmair, H.; Weber, E. R. *Phys. Rev. Lett.* 1998, *81*, 1243-1246.
[27] Hall, R. N.; Racette, J. H. *J. Appl. Phys.* 1964, *35*, 379-397.
[28] Estreicher, S. K.; Hastings, J. L. *Mater. Sci. Eng. B* 1999, *58*, 155-158.
[29] Weber, E. R. *Appl. Phys. A* 1983, *30*, 1-22.
[30] Keller, R.; Deicher, M.; Pfeiffer, W.; Skudlik, H.; Steiner, D.; Wichert, Th. *Phys. Rev. Lett.* 1990, *65*, 2023-2026.
[31] Aboelfotoh, M. O.; Svensson, B. G. *Phys. Rev. B* 1991, *44*, 12742-12747.
[32] Stolt, L.; d'Heurle, F. M. *Thin Solid Films* 1990, *189*, 269-274.
[33] Cros, A.; Aboelfotoh, M. O.; Tu, K. N. *J. Appl. Phys.* 1990, *67*, 3328-3336.
[34] Gas, P.; d'Heurle, F. M. *Appl. Surf. Sci.* 1993, *73*, 153-161.
[35] Veer, F. A.; Kolster, B. H.; Burgers, W. G. *Trans. Metall. Soc. AIME* 1968, *242*, 669-673.

[36] Mukherjee, K. P.; Bandyopadhyaya, J.; Gupta, K. P. *Trans. Metall. Soc. AIME* 1969, *245*, 2335-2338.
[37] Solberg, J. K. *Acta Cryst. A* 1978, *34*, 684-698.
[38] Weber, G.; Gillot, B.; Barret, P. *Phys. Stat. Sol. A* 1983, *75*, 567-576.
[39] Mundschau, M.; Bauer, E.; Telieps, W.; Swiech, W. *J. Appl. Phys.* 1989, *65*, 4747-4752.
[40] Kajbaji, M. E.; Thibault, J. *Phil. Mag. Lett.* 1995, *71*, 335-339.
[41] Harper, J. M. E.; Charai, A.; Stolt, L.; d'Heurle, F. M.; Fryer, P. M. *Appl. Phys. Lett.* 1990, *56*, 2519-2521.
[42] Stolt, L.; Charai, A.; d'Heurle, F. M.; Fryer, P. M.; Harper, J. M. E. *J. Vac. Sci. Technol. A* 1991, *9*, 1501-1505.
[43] Huang, H. Y.; Chen, L. J. *Mat. Res. Soc. Symp. Proc.* 1999, *564*, 217-222.
[44] Arcot, B.; Shy, Y. T.; Murarka, S. P.; Shepard, C.; Lanford, W. A. *Mat. Res. Soc. Symp. Proc.* 1991, *203*, 27-32.
[45] Y. T. Shy, S. P. Murarka, K. Singh, H. G. Bhimnathwala, J. M Borrego, C. Shepard, W. A. Lanford, *Minority carrier lifetime measurements of copper diffused through silicon dioxide in silicon by microwave reflection*, in: V. V. S. Rana, R. V. Joshi, I. Ohdomari (Eds.), Advanced Metallization Conference 1991, Murray Hill, NJ, U.S.A., October 8-10, 1991, Materials Research Society Conference Proceedings (1992) 433-436.
[46] Palleau, J.; Oberlin, J. C.; Braud, F.; Torres, J.; Mermet, J. L.; Mouche, M.-J.; Ermolieff, A.; Piaget, J. *Mat. Res. Soc. Symp. Proc.* 1994, *337*, 225-230.
[47] Miyazaki, H.; Kojima, H.; Hinode, K. *J. Appl. Phys.* 1997, *81*, 7746-7750.
[48] McBrayer, J. D.; Swanson, R. M.; Sigmon, T. W. *J. Electrochem. Soc.* 1986, *133*, 1242-1246.
[49] Pai, P.-L.; Ting, C. H. *IEEE Electron Device Lett.* 1989, *10*, 423-425.
[50] Vogt, M.; Drescher, K. *Appl. Surf. Sci.* 1995, *91*, 303-307.
[51] Raghavan, G.; Chiang, C.; Anders, P. B.; Tzeng, S.-M.; Villasol, R.; Bai, G.; Bohr, M.; Fraser, D. B. *Thin Solid Films* 1995, *262*, 168-176.
[52] Shacham-Diamand, Y.; Dedhia, A.; Hoffstetter, D.; Oldham, W. G. *J. Electrochem. Soc.* 1993, *140*, 2427-2432.
[53] Wendt, H.; Cerva, H.; Lehmann, V.; Pamler, W. *J. Appl. Phys.* 1989, *65*, 2402-2405.
[54] Movchan, B. A.; Demchishin, A. V. *Phys. Metals Metallogr.* 1969, *28*, 83-90.
[55] Thornton, J. A. *Ann. Rev. Mater. Sci.* 1977, *7*, 239-260.
[56] Thompson, C. V. *Ann. Rev. Mater. Sci.* 2000, *30*, 159-190.
[57] Lyman, T. *Metals Handbook;* 8th edition; American Society for Metals: Metals Park, OH, 1961; Vol. 1, p 30.

[58] Ezer, Y.; Härkönen, J.; Sokolov, V.; Saarilahti, J.; Kaitila, J.; Kuivalainen, P. *Mater. Res. Bull.* 1998, *33*, 1331-1337.
[59] Braud, F.; Torres, J.; Palleau, J.; Mermet, J. L.; Mouche, M. *J. Appl. Surf. Sci.* 1995, *91*, 251-256.
[60] Chuang, J.-C.; Tu, S.-L.; Chen, M.-C. *Thin Solid Films* 1999, *346*, 299-306.
[61] Mercier, M.; Weber, S.; Jacques, A.; Hirabayashi, H.; Ohkawa, H.; Kinoshita, M. *Defect Diffus. Forum* 1997, *143-147*, 1285-1290.
[62] Laurila, T.; Zeng, K.; Kivilahti, J. K.; Molarius, J.; Suni, I. *J. Appl. Phys.* 2000, *88*, 3377-3384.
[63] Ono, H.; Nakano, T.; Ohta, T. *Appl. Phys. Lett.* 1994, *64*, 1511-1513.
[64] Wang, S.-Q.; Suthar, S.; Hoeflich, C.; Burrow, B. J. *J. Appl. Phys.* 1993, *73*, 2301-2320.
[65] Fang, J. S.; Hsu, T. P.; Chen, G. S. *J. Electron. Mater.* 2006, *35*, 15-21.
[66] Noya, A.; Takeyama, M. B.; Sase, T. *J. Vac. Sci. Technol. B* 2005, *23*, 280-287.
[67] Thomas, R. E.; Guo, K. J.; Aaron, D. B.; Dobisz, E. A.; Perepezko, J. H.; Wiley, J. D. *Thin Solid Films* 1987, *150*, 245-252.
[68] Takeyama, M.; Kagomi, S.; Noya, A.; Sakanishi, K.; Sasaki, K. *J. Appl. Phys.* 1996, *80*, 569-573.
[69] Reid, J. S.; Kolawa, E.; Ruiz, R. P.; Nicolet, M.-A. *Thin Solid Films* 1993, *236*, 319-324.
[70] Wang, M. T.; Lin, Y. C.; Lee, J. Y.; Wang, C. C.; Chen, M. C. *J. Electrochem. Soc.* 1999, *146*, 1583-1592.
[71] Lee, Y.-J.; Suh, B.-S.; Park, C.-O. *Thin Solid Films* 1999, *357*, 237-241.
[72] Riedel, S.; Schulz, S. E.; Baumann, J.; Rennau, M.; Gessner, T. *Microelectron. Eng.* 2001, *55*, 213-218.
[73] Takeyama, M. B.; Itoi, T.; Satoh, K.; Sakagami, M.; Noya, A. *J. Vac. Sci. Technol. B* 2004, *22*, 2542-2547.
[74] Chen, C.-S.; Liu, C.-P.; Yang, H.-G.; Tsao, C. Y. A. *J. Vac. Sci. Technol. B* 2004, *22*, 1075-1083.
[75] Alén, P.; Ritala, M.; Arstila, K.; Keinonen, J.; Leskelä, M. *Thin Solid Films* 2005, *491*, 235-241.
[76] Ou, K. L. *Microelectron. Eng.* 2006, *83*, 312-318.
[77] Hecker, M.; Hübner, R.; Ecke, R.; Schulz, S.; Engelmann, H.-J.; Stegmann, H.; Hoffmann, V.; Mattern, N.; Gessner, T.; Zschech, E. *Microelectron. Eng.* 2002, *64*, 269-277.
[78] Hecker, M.; Fischer, D.; Hoffmann, V.; Engelmann, H.-J.; Voss, A.; Mattern, N.; Wenzel, C.; Vogt, C.; Zschech E. *Thin Solid Films* 2002, *414*, 184-191.

[79] Wang, S. J.; Tsai, H. Y.; Sun, S. C.; Shiao, M. H. *J. Electrochem. Soc.* 2001, *148*, G500-G506.
[80] Tsai, H. Y.; Sun, S. C.; Wang, S. J. *J. Electrochem. Soc.* 2000, *147*, 2766-2772.
[81] Charai, A.; Hörnström, S. E.; Thomas, O.; Fryer, P. M.; Harper, J. M. E. *J. Vac. Sci. Technol. A* 1989, *7*, 784-789.
[82] Hartman, J. W.; Yeh, H.; Atwater, H. A.; Hashim, I. *Mat. Res. Soc. Symp. Proc.* 1999, *564*, 257-262.
[83] Pelleg, J.; Sade, G. *J. Appl. Phys.* 2002, *91*, 6099-6104.
[84] No, J.-T.; O, J.-H.; Lee, C. *Mater. Chem. Phys.* 2000, *63*, 44-49.
[85] Hecker, M.; Hübner, R.; Acker, J.; Hoffmann, V.; Mattern, N.; Ecke, R.; Schulz, S.E.; Heuer, H.; Wenzel, C.; Engelmann, H.-J.; Zschech, E. In *Materials for Information Technology – Devices, Interconnects and Packaging;* Zschech, E.; Whelan, C.; Mikolajick, T.; Eds.; Springer-Verlag London Limited, 2005; pp 283-295.
[86] Lai, L. W.; Chen, J. S.; Hsu, W.-S. *J. Appl. Phys.* 2003, *94*, 5396-5398.
[87] Liu, Y.; Song, S.; Mao, D.; Ling, H.; Li, M. *Microelectron. Eng.* 2004, *75*, 309-315.
[88] H. Fang, T. Weidman, A. Shanmugasundram, M. Naik, B. Kapoor, *Electroless Co(W,P) capping application development*, in: D. Erb (Ed.), Advanced Metallization Conference 2004, San Diego, CA, U.S.A., October 19-21, 2004, Materials Research Society Conference Proceedings (2005) 849-852.
[89] Yang, L. C.; Hsu, C. S.; Chen, G. S.; Fu, C. C.; Zuo, J. M.; Lee, B. Q. *Appl. Phys. Lett.* 2005, *87*, 121911/1-121911/3.
[90] Reid, J. S.; Liu, R. Y.; Smith, P. M.; Ruiz, R. P.; Nicolet, M.-A. *Thin Solid Films* 1995, *262*, 218-223.
[91] Koh, W.; Kumar, D.; Li, W.-M.; Sprey, H.; Raaijmakers, I. *J. Solid State Technol.* 2005, *48*, 54-56, 58.
[92] Rawal, S.; Norton, D. P.; Anderson, T. J.; McElwee-White, L. *Appl. Phys. Lett.* 2005, *87*, 111902/1-111902/3.
[93] Chen, C.-S.; Liu, C.-P. *J. Non-Cryst. Solids* 2005, *351*, 3725-3729.
[94] Kim, H. C.; Alford, T. L. *Thin Solid Films* 2004, *449*, 6-11.
[95] S. Lopatin, A. Shanmugasundram, D. Lubomirsky, I. A. Pancham, *Method and tool of chemical doping Co-W alloys with Re for increasing barrier properties of electroless capping layers for IC Cu interconnects*, U.S. Pat. Appl. Publ. (2005), 17 pp. CODEN: USXXCO US 20050101130 A1 20050512 Patent written in English. Application: US 2003-703769 20031107.

[96] Senkevich, J. J.; Wang, P.-I.; Wiegand, C. J.; Lu, T.-M. *Appl. Phys. Lett.* 2004, *84*, 2617-2619.
[97] Ganesan, P. G.; Singh, A. P.; Ramanath, G. *Appl. Phys. Lett.* 2004, *85*, 579-581.
[98] Chin, B. L.; Yao, G.; Ding, P.; Fu, J.; Chen, L. *Semiconductor International* 2001, *24(5)*, 107-108, 110, 112, 114.
[99] Lane, M. W.; Murray, C. E.; McFeely, F. R.; Vereecken, P. M.; Rosenberg, R. *Appl. Phys. Lett.* 2003, *83*, 2330-2332.
[100] Kim, H.; Koseki, T.; Ohba, T.; Ohta, T.; Kojima, Y.; Sato, H.; Shimogaki, Y. *J. Electrochem. Soc.* 2005, *152*, G594-G600.
[101] Subramanian, P. R.; Laughlin, D. E. In *Binary alloy phase diagrams*; 2nd Edition; Massalski, T. B.; Okamoto, H.; Subramanian, P. R.; Kacprazak, L.; Eds.; ASM International: Materials Park, OH, 1990; Vol. 2, p 1467.
[102] Chyan, O.; Arunagiri, T. N.; Ponnuswamy, T. *J. Electrochem. Soc.* 2003, *150*, C347-C350.
[103] Chan, R.; Arunagiri, T. N.; Zhang, Y.; Chyan, O.; Wallace, R. M.; Kim, M. J.; Hurd, T. Q. *Electrochem. Solid-State Lett.* 2004, *7*, G154-G157.
[104] Josell, D.; Wheeler, D.; Witt, C.; Moffat, T. P. *Electrochem. Solid-State Lett.* 2003, *6*, C143-C145.
[105] Arunagiri, T. N.; Zhang, Y.; Chyan, O.; El-Bouanani, M.; Kim, M. J.; Chen, K. H.; Wu, C. T.; Chen, L. C. *Appl. Phys. Lett.* 2005, *86*, 083104/1-083104/3.
[106] Arunagiri, T. N.; Zhang, Y.; Chyan, O.; Kim, M. J.; Hurd, T. Q. *J. Electrochem. Soc.* 2005, *152*, G808-G812.
[107] Damayanti, M.; Sritharan, T.; Mhaisalkar, S. G.; Gan, Z. H. *Appl. Phys. Lett.* 2006, *88*, 044101/1-044101/3.
[108] Josell, D.; Bonevich, J. E.; Moffat, T. P.; Aaltonen, T.; Ritala, M.; Leskelä M. *Electrochem. Solid-State Lett.* 2006, *9*, C48-C50.
[109] Tsukimoto, S.; Morita, T.; Moriyama, M.; Ito, K.; Murakami, M. *J. Electron. Mater.* 2005, *34*, 592-599.
[110] Chu, J. P.; Lin, C. H. *Appl. Phys. Lett.* 2005, *87*, 211902/1-211902/3.
[111] Koike, J.; Wada, M. *Appl. Phys. Lett.* 2005, *87*, 041911/1-041911/3.
[112] Forster, J. In *Ionized Physical Vapor Deposition*; Hopwood, J. A.; Ed.; Academic Press: New York, NY, 2000; p 141.
[113] Gopalraja, P.; Forster, J. *Appl. Phys. Lett.* 2000, *77*, 3526-3528.
[114] Li, N.; Ruzic, D. N.; Powell, R. A. *J. Vac. Sci. Technol. B* 2004, *22*, 2237-2242.
[115] Kim, H. *Surf. Coat. Technol.* 2006, *200*, 3104-3111.

[116] Alén, P.; Ritala, M.; Arstila, K.; Keinonen, J.; Leskelä, M. *J. Electrochem. Soc.* 2005, *152*, G361-G366.
[117] Cheng, H.-E.; Lee, W.-J.; Hsu, C.-M. *Thin Solid Films* 2005, *485*, 59-65.
[118] Bystrova, S.; Aarnink, A. A. I.; Holleman, J.; Wolters, R. A. M. *J. Electrochem. Soc.* 2005, *152*, G522-G527.
[119] Kim, H.; Cabral, Jr., C.; Lavoie, C.; Rossnagel, S. M. *J. Vac. Sci. Technol. B* 2002, *20*, 1321-1326.
[120] Leskelä, M.; Ritala, M. *J. Phys. IV* 1999, *9-Pr8*, 837-852.
[121] Kim, H.; Kellock, A. J.; Rossnagel, S. M. *J. Appl. Phys.* 2002, *92*, 7080-7085.
[122] Park, J.-S.; Lee, M.-J.; Lee, C.-S.; Kang, S.-W. *Electrochem. Solid-State Lett.* 2001, *4*, C17-C19.
[123] Park, J.-S.; Park, H.-S.; Kang, S.-W. *J. Electrochem. Soc.* 2002, *149*, C28-C32.
[124] Aaltonen, T.; Alén, P.; Ritala, M.; Leskelä, M. *Chem. Vap. Depos.* 2003, *9*, 45-49.
[125] Min, Y.-S.; Bae, E. J.; Jeong, K. S.; Cho, Y. J.; Lee, J.-H.; Choi, W. B.; Park, G.-S. *Adv. Mater.* 2003, *15*, 1019-1022.
[126] Kwon, O.-K.; Kwon, S.-H.; Park, H.-S.; Kang, S.-W. *J. Electrochem. Soc.* 2004, *151*, C753-C756.
[127] Subramanian, P. R.; Laughlin, D. E. *Bull. Alloy Phase Diagrams* 1989, *10*, 652-655.
[128] Kaufman, L. *Calphad* 1991, *15*, 243-259.
[129] Zier, M.; Oswald, S.; Reiche, R.; Kozlowska, M.; Wetzig, K. *J. Elec. Spec. Relat. Phenom.* 2004, *137–140*, 229–233.
[130] Cheng, J. Y.; Chen, L. J. *J. Appl. Phys.* 1991, *69*, 2161-2168.
[131] Bevelo, A. J.; Campisi, G. J.; Shanks, H. R.; Schmidt, F. A. *J. Appl. Phys.* 1980, *51*, 5390-5395.
[132] Chen, L.; Ekstrom, B.; Kelber, J. *Mat. Res. Soc. Symp. Proc.* 1999, *564*, 287-292.
[133] Zong, Y.; Shan, X.; Watkins, J. J. *Langmuir* 2004, *20*, 9210-9216.
[134] Lane, M.; Dauskardt, R. H.; Krishna, N.; Hashim, I. *J. Mater. Res.* 2000, 15, 203-211.
[135] Mueller, M. H. *Scr. Metall.* 1977, *11*, 693.
[136] Smirnov, Yu. M.; Finkel, V. A. *Soviet Physics JETP* 1966, *22*, 750-753.
[137] Baker, P. N. *Thin Solid Films* 1972, *14*, 3-25.
[138] Arakcheeva, A.; Chapuis, G.; Grinevitch, V. *Acta Cryst. B* 2002, *58*, 1-7.
[139] Liu, L.; Wang, Y.; Gong, H. *J. Appl. Phys.* 2001, *90*, 416-420.
[140] Holloway, K.; Fryer, P. M. *Appl. Phys. Lett.* 1990, *57*, 1736-1738.

[141] Fischer, D.; Meissner, O.; Bendjus, B.; Schreiber, J.; Stavrev, M.; Wenzel, C. *Surf. Interface Anal.* 1997, *25*, 522-528.
[142] Face, D. W.; Prober, D. E. *J. Vac. Sci. Technol. A* 1987, *5*, 3408-3411.
[143] Chen, G. S.; Lee, P. Y.; Chen, S. T. *Thin Solid Films* 1999, *353*, 264-273.
[144] Mutscheller, A.; Clevenger, L. A.; Harper, J. M. E.; Cabral, Jr., C.; Barmak, K. *Mat. Res. Soc. Symp. Proc.* 1992, *239*, 51-56.
[145] Clevenger, L. A.; Mutscheller, A.; Harper, J. M. E.; Cabral, Jr., C.; Barmak, K. *J. Appl. Phys.* 1992, *72*, 4918-4924.
[146] Hoogeveen, R.; Moske, M.; Geisler, H.; Samwer, K. *Thin Solid Films* 1996, *275*, 203-206.
[147] Moshfegh, A. Z.; Akhavan, O. *Thin Solid Films* 2000, *370*, 10-17.
[148] Stavrev, M.; Fischer, D.; Praessler, F.; Wenzel, C.; Drescher, K. *J. Vac. Sci. Technol. A* 1999, *17*, 993-1001.
[149] Catania, P.; Doyle, J. P.; Cuomo, J. J. *J. Vac. Sci. Technol. A* 1992, *10*, 3318-3321.
[150] Holloway, K.; Fryer, P. M.; Cabral, Jr., C.; Harper, J. M. E.; Bailey, P. J.; Kelleher, K. H. *J. Appl. Phys.* 1992, *71*, 5433-5444.
[151] Clevenger, L. A.; Bojarczuk, N. A.; Holloway, K.; Harper, J. M. E.; Cabral, Jr., C.; Schad, R. G.; Cardone, F.; Stolt, L. *J. Appl. Phys.* 1993, *73*, 300-308.
[152] Yin, K.-M.; Chang, L.; Chen, F.-R.; Kai, J.-J.; Chiang, C.-C.; Chuang, G.; Ding, P.; Chin, B.; Zhang, H.; Chen, F. *Thin Solid Films* 2001, *388*, 27-33.
[153] Laurila, T.; Zeng, K.; Kivilahti, J. K.; Molarius, J.; Suni, I. *J. Mater. Res.* 2001, *16*, 2939-2946.
[154] Kang, B.-S.; Lee, S.-M.; Kwak, J. S.; Yoon, D.-S.; Baik, H. K. *J. Electrochem. Soc.* 1997, *144*, 1807-1812.
[155] Yoon, D.-S.; Baik, H. K.; Lee, S.-M. *J. Appl. Phys.* 1998, *83*, 1333-1336.
[156] Kwak, J. S.; Baik, H. K.; Kim, J.-H.; Lee, S.-M. *Appl. Phys. Lett.* 1998, *72*, 2832-2834.
[157] Kwak, J. S.; Baik, H. K.; Kim, J.-H.; Lee, S.-M.; Ryu, H. J.; Je, J. H. *J. Appl. Phys.* 1999, *85*, 6898-6903.
[158] C.-K. Hu, S. Chang, M. B. Small, J. E. Lewis, *Diffusion barrier studies for Cu*, Proceedings of the third International VLSI Multilevel Interconnection Conference, June 9-10, 1986, Santa Clara, CA, U.S.A., Institute of Electrical and Electronics Engineers (IEEE) (1986) 181-187.
[159] Olowolafe, J. O.; Mogab, C. J.; Gregory, R. B. *Thin Solid Films* 1993, *227*, 37-43.
[160] Wang, M. T.; Lin, Y. C.; Chen, M. C. *J. Electrochem. Soc.* 1998, *145*, 2538-2545.

[161] Jang, S.-Y.; Lee, S.-M.; Baik, H.-K. *J. Mater. Sci.: Mater. Electron.* 1996, *7*, 271-278.
[162] Maex, K.; Baklanov, M. R.; Shamiryan, D.; Iacopi, F.; Brongersma, S. H.; Yanovitskaya, Z. S. *J. Appl. Phys.* 2003, *93*, 8793-8841.
[163] Chen, Z.; Prasad, K.; Li, C. Y.; Lu, P. W.; Su, S. S.; Tang, L. J.; Gui, D.; Balakumar, S.; Shu, R.; Kumar, R. *Appl. Phys. Lett.* 2004, *84*, 2442-2444.
[164] Chen, Z.; Prasad, K.; Jiang, N.; Tang, L. J.; Lu, P. W.; Li, C. Y. *J. Vac. Sci. Technol. B* 2005, *23*, 1866-1872.
[165] Gupta, D. *Mater. Chem. Phys.* 1995, *41*, 199-205.
[166] Sun, X.; Kolawa, E.; Chen, J.-S.; Reid, J. S.; Nicolet, M.-A. *Thin Solid Films* 1993, *236*, 347-351.
[167] Stavrev, M.; Fischer, D.; Wenzel, C.; Drescher, K.; Mattern, N. *Thin Solid Films* 1997, *307*, 79-88.
[168] J. Baumann, G. Schwalbe, S. Zimmermann, C. Kaufmann, T. Gessner, *Phase, structure and properties of sputtered Ta and TaN_x films*, in: A. J. McKerrow, Y. Shacham-Diamand, S. Zaima, T. Ohba (Eds.), Advanced Metallization Conference 2001, Montreal, Canada, October 8-11, 2001, Materials Research Society Conference Proceedings (2002) 613-618.
[169] Shin, C.-S.; Kim, Y.-W.; Gall, D.; Greene, J. E.; Petrov, I. *Thin Solid Films* 2002, *402*, 172-182.
[170] Mashimo, T.; Tashiro, S.; Toya, T.; Nishida, M.; Yamazaki, H.; Yamaya, S.; Oh-Ishi, K.; Syono, Y. *J. Mater. Sci.* 1993, *28*, 3439-3443.
[171] Greenwood, N. N.; Earnshaw, A. *Chemistry of the elements;* 2nd edition; Butterworth-Heinemann: Oxford, 1997; p 418.
[172] Ritala, M.; Kalsi, P.; Riihelä, D.; Kukli, K.; Leskelä, M.; Jokinen, *J. Chem. Mater.* 1999, *11*, 1712-1718.
[173] Alén, P.; Juppo, M.; Ritala, M.; Sajavaara, T.; Keinonen, J.; Leskelä, M. *J. Electrochem. Soc.* 2001, *148*, G566-G571.
[174] Juppo, M.; Ritala, M.; Leskelä, M. *J. Electrochem. Soc.* 2000, *147*, 3377-3381.
[175] Alén, P.; Juppo, M.; Ritala, M.; Leskelä, M.; Sajavaara, T.; Keinonen, J. *J. Mater. Res.* 2002, *17*, 107-114.
[176] Burgess, S. R.; Donohue, H.; Buchanan, K.; Rimmer, N.; Rich, P. *Microelectron. Eng.* 2002, *64*, 307-313.
[177] S. Zimmermann, J. Baumann, C. Kaufmann, T. Gessner, *Thermal stability of thin Ta and TaN_x films as diffusion barriers for copper metallization*, in: B. M. Melnick, T. S. Cale, S. Zaima, T. Ohta (Eds.), Advanced Metallization Conference 2002, San Diego, California, U.S.A., October 1-3, 2002, Materials Research Society Conference Proceedings (2003) 859-864.

[178] Kim, H.; Lavoie, C.; Copel, M.; Narayanan, V.; Park, D.-G.; Rossnagel, S. M. *J. Appl. Phys.* 2004, *95*, 5848-5855.
[179] Min, K.-H.; Chun, K.-C.; Kim, K.-B. *J. Vac. Sci. Technol. B* 1996, *14*, 3263-3269.
[180] Chen, G. S.; Chen, S. T. *J. Appl. Phys.* 2000, *87*, 8473-8482.
[181] Oku, T.; Kawakami, E.; Uekubo, M.; Takahiro, K.; Yamaguchi, S.; Murakami, M. *Appl. Surf. Sci.* 1996, *99*, 265-272.
[182] Lin, J.-C.; Lee, C. *J. Electrochem. Soc.* 1999, *146*, 3466-3471.
[183] T. Nogami, Y.-C. Joo, S. Lopatin, J. Romero, J. Bernard, W. Blum, H.-J. Engelmann, J. Gray, B. Tracy, S. Chen, T. Lukanc, D. Brown, P. Besser, G. Morales, R. Cheung, *Graded Ta/TaN/Ta barrier for copper interconnects for high electromigration resistance*, in: G. S. Sandhu, H. Koerner, M. Murakami, Y. Yasuda, N. Kobayashi (Eds.), Advanced Metallization Conference 1998, Colorado Springs, U.S.A., October 6-8, 1998, Materials Research Society Conference Proceedings (1999) 313-319.
[184] Chen, G. S.; Huang, S. C.; Chen, S. T.; Yang, T. J.; Lee, P. Y.; Jou, J. H.; Lin, T. C. *Appl. Phys. Lett.* 2000, *76*, 2895-2897.
[185] D. Edelstein, C. Uzoh, C. Cabral, Jr., P. DeHaven, P. Buchwalter, A. Simon, E. Cooney, S. Malhotra, D. Klaus, H. Rathore, B. Agarwala, D. Nguyen, *An optimal liner for copper damascene interconnects*, in: A. J. McKerrow, Y. Shacham-Diamand, S. Zaima, T. Ohba (Eds.), Advanced Metallization Conference 2001, Montreal, Canada, October 8-11, 2001, Materials Research Society Conference Proceedings (2002) 541-547.
[186] K. Ishikawa, M. Miyauchi, H. Ashihara, T. Saitoh, U. Tanaka, T. Ohshima, K. Torii, S. Ishihara, H. Aoki, H. Yamaguchi, K. Hinode, T. Saito, *Electromigration resistance improvement of dual-damascene copper interconnection using TaN/Ta barrier formed by ionized bias sputtering*, in: A. J. McKerrow, Y. Shacham-Diamand, S. Zaima, T. Ohba (Eds.), Advanced Metallization Conference 2001, Montreal, Canada, October 8-11, 2001, Materials Research Society Conference Proceedings (2002) 487-492.
[187] Traving, M.; Zienert, I.; Zschech, E.; Schindler, G.; Steinhögl, W.; Engelhardt, M. *Appl. Surf. Sci.* 2005, *252*, 11-17.
[188] Ho, P.; Rajagopalan, R.; Chong, H.; Chung, H.; Yu, J. *Semiconductor International* 2004, *27(6)*, 61-62, 64, 67.
[189] Murarka, S. P.; Fraser, D. B. *J. Appl. Phys.* 1980, *51*, 1593-1598.
[190] Baiocchi, F. A.; Lifshitz, N.; Sheng, T. T.; Murarka, S. P. *J. Appl. Phys.* 1988, *64*, 6490-6495.
[191] D. Fischer, M. Stavrev, N. Urbansky, C. Wenzel, E. Neubauer, J. G. Bauer, T. Scherg, H.-J. Schulze, *Barrier and contact behavior of tantalum-based*

thin films for use in copper metallization scheme, in: G. S. Sandhu, H. Koerner, M. Murakami, Y. Yasuda, N. Kobayashi (Eds.), Advanced Metallization Conference 1998, Colorado Springs, U.S.A., October 6-8, 1998, Materials Research Society Conference Proceedings (1999) 337-344.

[192] Kolawa, E.; Chen, J. S.; Reid, J. S.; Pokela, P. J.; Nicolet, M.-A. *J. Appl. Phys.* 1991, *70*, 1369-1373.

[193] Oizumi, M.; Aoki, K.; Hashimoto, S.; Nemoto, S.; Fukuda, Y. *Jpn. J. Appl. Phys.* 2000, *39*, 1291-1294.

[194] Bicker, M.; Pinnow, C.-U.; Geyer, U.; Schneider, S.; Seibt, M. *Appl. Phys. Lett.* 2001, *78*, 3618-3620.

[195] Suh, Y.-S.; Heuss, G. P.; Misra, V.; Park, D.-G.; Lim, K.-Y. *J. Electrochem. Soc.* 2003, *150*, F79-F82.

[196] Pokela, P. J.; Reid, J. S.; Kwok, C.K.; Kolawa, E.; Nicolet, M.-A. *J. Appl. Phys.* 1991, *70*, 2828-2832.

[197] Cabral, Jr., C.; Saenger, K. L.; Kotecki, D. E.; Harper, J. M. E. *J. Mater. Res.* 2000, *15*, 194-198.

[198] Kwok, C.-K.; Kolawa, E.; Nicolet, M.-A.; Lee, R. C. *Mat. Res. Soc. Symp. Proc.* 1991, *226*, 261-266.

[199] R. Venkatraman, J. Mendonca, T. P. Ong, G. Hamilton, B. Rogers, L. Frisa, C. Simpson, V. Kaushik, R. Gregory, E. Apen, D. Coronell, M. Angyal, A. Jain, M. Herrick, C. Capasso, R. Fiordalice, J. Klein, E. Weitzman, *Process development and integration of PVD Ta-Si-N films for copper interconnect applications*, in: R. Cheung, J. Klein, K. Tsubouchi, M. Murakami, N. Kobayashi (Eds.), Advanced Metallization Conference 1997, San Diego, California, U.S.A., September 30 - October 2, 1997, Materials Research Society Conference Proceedings (1998) 63-70.

[200] Hara, T.; Yoshida, Y.; Toida, H. *Electrochem. Solid State Lett.* 2002, *5*, G36-G39.

[201] Lee, H.; Lopatin, S. D. *Thin Solid Films* 2005, *492*, 279-284.

[202] Kolawa, E.; Pokela, P. J.; Reid, J. S.; Chen, J. S.; Ruiz, R. P.; Nicolet, M.-A. *IEEE Electron Device Lett.* 1991, *12*, 321-323.

[203] Lin, C.-L.; Ku, S.-R.; Chen, M.-C. *Jpn. J. Appl. Phys.* 2001, *40*, 4181-4186.

[204] Cabral, Jr., C.; Lavoie, C.; Harper, J. M. E.; Jordan-Sweet, J. *Thin Solid Films* 2001, *397*, 194-202.

[205] Aouadi, S. M.; Zhang, Y.; Bohnhoff, A.; Lee, J.; Williams, M. *J. Vac. Sci. Technol. A* 2005, *23*, 1359-1363.

[206] Angyal, M. S.; Shacham-Diamand, Y.; Reid, J. S.; Nicolet, M.-A. *Appl. Phys. Lett.* 1995, *67*, 2152-2154.

[207] Lee, Y.-J.; Suh, B.-S.; Kwon, M. S.; Park, C.-O. *J. Appl. Phys.* 1999, *85*, 1927-1934.
[208] Fischer, D.; Scherg, T.; Bauer, J. G.; Schulze, H.-J.; Wenzel, C. *Microelectron. Eng.* 2000, *50*, 459-464.
[209] Windt, D. L. *Comput. Phys.* 1998, *12*, 360-370.
[210] *Powder Diffraction File*, Joint Committee on Powder Diffraction Standards; International Centre for Diffraction Data: Newtown Square, PA, 2001.
[211] Hoffmann, V.; Uhlemann, H.-J.; Präßler, F.; Wetzig, K.; Birus, D. *Fresenius J. Anal. Chem.* 1996, *355*, 826-830.
[212] Hoffmann, V.; Hecker, M.; Hübner, R. *Z. Metallkd.* 2005, *96*, 983-987.
[213] Williams, D. B.; Carter C. B. *Transmission Electron Microscopy;* Plenum Press: New York, NY, 1996.
[214] Klug, H. P.; Alexander, L. E. *X-ray diffraction procedures. For polycrystalline and amorphous materials.;* John Wiley & Sons, Inc.: New York, NY; Chapman & Hall, Limited: London, 1954.
[215] Ikeda, M.; Murooka, M.; Suzuki, K. *Jpn. J. Appl. Phys.* 2002, *41*, 3902-3908.
[216] Kim, D. J.; Kim, Y. T.; Park, J.-W. *J. Appl. Phys.* 1997, *82*, 4847-4851.
[217] Nötzold, K. Diploma Thesis; Zwickau University of Applied Sciences, 2003.
[218] Fischer, D. Diploma Thesis; Dresden University of Technology, 2002.
[219] Stoney, G. *Proc. R. Soc. London A* 1909, *82*, 172-175.
[220] Hübner, R. Hecker, M. Mattern, N. Hoffmann, V. Wetzig, K. Engelmann, H.-J. Zschech, E. *Anal. Bioanal. Chem.* 2004, *379*, 568-575.
[221] Cabral, Jr., C.; Clevenger, L. A.; Schad, R. G. *J. Vac. Sci. Technol. B* 1994, *12*, 2818-2821.
[222] Samsonova, G. V. *Fiziko-chimiceskie svojstva okislov;* Metallurgija: Moskva 1978; p 148.
[223] Powers, R. W.; Doyle, M. V. *J. Appl. Phys.* 1959, *30*, 514-524.
[224] Cros, A.; Tu, K. N. *J. Appl. Phys.* 1986, *60*, 3323-3326.
[225] Giber, J.; Oechsner, H. *Thin Solid Films* 1985, *131*, 279-287.
[226] Kalinovich, D. F.; Kovenskii, I. I.; Smolin, M. D. *Phys. Metalls Metallog. [USSR]* 1964, *18*, 154-155.
[227] Laurila, T.; Zeng, K.; Kivilahti, J. K.; Molarius, J.; Riekkinen, T.; Suni, I. *Microelectron. Eng.* 2002, *60*, 71-80.
[228] Pretorius, R.; Harris, J. M.; Nicolet, M.-A. *Solid State Electron.* 1978, *21*, 667-675.
[229] Beyers, R. *J. Appl. Phys.* 1984, *56*, 147-152.

[230] Brebec, G.; Seguin, R.; Sella, C.; Bevenot, J.; Martin, J. C. *Acta Metall.* 1980, *28*, 327-333.
[231] Fontbonne, A.; Gilles, J.-C. *Revue Internationale des Hautes Temperatures et des Refractaires* 1969, *6*, 181-192.
[232] Hübner, R.; Hecker, M.; Mattern, N.; Hoffmann, V.; Wetzig, K.; Wenger, Ch.; Engelmann, H.-J.; Wenzel, Ch.; Zschech, E.; Bartha, J. W. *Thin Solid Films* 2003, *437*, 248-256.
[233] Jiran, E.; Thompson, C. V. *J. Electron. Mater.* 1990, *19*, 1153-1160.
[234] Mullins, W. W. *J. Appl. Phys.* 1957, *28*, 333-339.
[235] Srolovitz, D. J.; Safran, S. A. *J. Appl. Phys.* 1986, *60*, 247-254.
[236] Knacke, O.; Kubaschewski, O.; Hesselmann K. *Thermochemical Properties of Inorganic Substances;* 2nd edition; Springer-Verlag: Berlin; Verlag Stahleisen m.b.H.: Düsseldorf, 1991; Vol. 1 and 2.
[237] Hübner, R.; Hecker, M.; Mattern, N.; Hoffmann, V.; Wetzig, K.; Wenger, Ch.; Engelmann, H.-J.; Wenzel, Ch.; Zschech, E. *Thin Solid Films* 2004, *458*, 237-245.
[238] Hübner, R.; Hecker, M.; Mattern, N.; Voss, A.; Acker, J.; Hoffmann, V.; Wetzig, K.; Engelmann, H.-J.; Zschech, E.; Heuer, H.; Wenzel, Ch. *Thin Solid Films* 2004, *468*, 183-192.
[239] Hübner, R.; Reiche, R.; Hecker, M.; Mattern, N.; Hoffmann, V.; Wetzig, K.; Heuer, H.; Wenzel, Ch.; Engelmann, H.-J.; Zschech, E. *Cryst. Res. Technol.* 2005, *40*, 135-142.
[240] Shepherd, K.; Kelber, J. *Appl. Surf. Sci.* 1999, *151*, 287-298.
[241] Kleber, W. *Einführung in die Kristallographie;* Verlag Technik: Berlin, 1998; p 204.
[242] Feltz, A. *Amorphe und glasartige anorganische Festkörper;* Akademie-Verlag: Berlin, 1983; p 66.
[243] Schlesinger, M. E. In *Binary alloy phase diagrams;* 2nd Edition; Massalski, T. B.; Okamoto, H.; Subramanian, P. R.; Kacprazak, L.; Eds.; ASM International: Materials Park, OH, 1990; Vol. 3, p 3364.
[244] Olesinski, R. W.; Abbaschian, G. J. In *Binary alloy phase diagrams;* 2nd Edition; Massalski, T. B.; Okamoto, H.; Subramanian, P. R.; Kacprazak, L.; Eds.; ASM International: Materials Park, OH, 1990; Vol. 2, p 1477.
[245] Hübner, R.; Hecker, M.; Mattern, N.; Hoffmann, V.; Wetzig, K.; Heuer, H.; Wenzel, Ch.; Engelmann, H.-J.; Gehre, D.; Zschech, E. *Thin Solid Films* 2006, *500*, 259-267.
[246] Nguyen Tan, T. A.; Azizan, M.; Derrien, J. *Surf. Sci.* 1987, *189/190*, 339-345.
[247] Walser, R. M.; Bené, R. W. *Appl. Phys. Lett.* 1976, *28*, 624-625.

[248] Noya, A.; Takeyama, M.; Sasaki, K.; Nakanishi, T. *J. Appl. Phys.* 1994, *76*, 3893-3895.
[249] Noya, A.; Takeyama, M.; Sasaki, K.; Aoyagi, E.; Hiraga, K. *J. Vac. Sci. Technol. A* 1997, *15*, 253-257.
[250] Bené, R. W. *J. Appl. Phys.* 1987, *61*, 1826-1833.
[251] Baglin, J. E. E.; d'Heurle, F. M.; Hammer, W. N.; Petersson, S. *Nucl. Instrum. Methods* 1980, *168*, 491-497.
[252] Tsaur, B. Y.; Lau, S. S.; Meyer, J. W.; Nicolet, M.-A. *Appl. Phys. Lett.* 1981, *38*, 922-924.
[253] Tu, K. N.; Mayer, J. W. In *Thin films – Interdiffusion and reactions;* Poate, J. M.; Tu, K. N.; Mayer J. W.; Eds.; John Wiley & Sons, Inc.: New York, NY, 1978; 359-405.
[254] Reid, J. S. PhD Thesis; California Institute of Technology: Pasadena, CA, 1995.
[255] Hübner, R.; Hecker, M; Wetzig, K.; Heuer, H.; Wenzel, Ch.; Engelmann, H.-J.; Zschech, E. In E. Zschech, K. Maex, P. S. Ho, H. Kawasaki, T. Nakamura (Eds.), Eighth International Workshop on Stress-Induced Phenomena in Metallization, Dresden, Germany, September 12-14, 2005, AIP Conference Proceedings 817 (2006) 59-64.

INDEX

A

AAS, 26, 35, 62, 63, 64, 76
acceptor, 16
acid, 20, 35
acrylic acid, 23
activation, 15, 16, 17, 19, 25, 27
activation energy, 15, 16, 17, 19, 25
activation enthalpy, 15
activation entropy, 15
adhesion, 10, 18, 20, 21, 23, 26, 27, 28
agent, 22, 26
agents, 26
AIP, 81, 94
air, 10, 20, 59
alloys, 19, 85
alternative, 20
aluminum, 7, 10
AMD, 79
amine, 19
amines, 26
ammonium, 35, 64
amorphous, 18, 19, 20, 23, 24, 25, 26, 27, 28, 32, 38, 41, 45, 46, 54, 56, 59, 60, 62, 65, 66, 67, 68, 69, 70, 72, 73, 75, 76, 92
amorphous carbon, 32
amplitude, 49, 57
analytical techniques, 8, 29, 33, 57, 62, 63, 66, 77, 78
annealing, 7, 8, 13, 17, 19, 20, 21, 23, 24, 25, 26, 28, 29, 31, 33, 43, 44, 45, 46, 47, 48, 49, 50, 51, 52, 53, 54, 55, 56, 57, 58, 60, 62, 63, 65, 66, 67, 68, 69, 70, 71, 72, 73, 74, 75, 76, 77, 78
application, 7, 10, 13, 16, 18, 19, 21, 23, 25, 27, 47, 54, 66, 78, 85
argon, 17, 26
aspect ratio, 21
atomic absorption spectrometry, 8, 26, 35, 63, 77
atoms, 11, 12, 15, 16, 18, 21, 24, 25, 26, 28, 41, 44, 47, 51, 53, 56, 57, 58, 62, 63, 64, 65, 66, 68, 69, 70, 73, 75, 76
Auger electron spectroscopy, 7
Aurora, 22

B

backscattering, 32
barrier, 7, 8, 10, 16, 18, 19, 20, 21, 22, 23, 24, 25, 26, 27, 28, 29, 31, 33, 34, 35, 37, 38, 39, 40, 41, 43, 44, 47, 49, 51, 52, 53, 54, 56, 57, 58, 60, 62, 63, 64, 65, 66, 68, 70, 71, 72, 73, 75, 76, 77, 78, 85, 88, 90
barriers, 7, 10, 18, 19, 20, 21, 22, 23, 24, 25, 26, 28, 29, 31, 33, 37, 40, 42, 47, 57, 59, 60, 63, 72, 74, 75, 76, 77, 78, 89
baths, 20
behavior, 8, 10, 17, 18, 28, 37, 45, 47, 52, 54, 56, 66, 67, 70, 71, 73, 76, 77, 90

bias, 7, 10, 17, 19, 21, 90
Bohr, 83
Boltzmann constant, 11
bonding, 20, 41
bonds, 41, 57, 68
boron, 16
boundary conditions, 13
breakdown, 25
broad spectrum, 34

C

California, 89, 91, 94
Canada, 89, 90
candidates, 19, 25, 29
capacitance, 9
carbide, 25
carbon, 22, 34, 44
carboxyl, 19
carrier, 22, 83
cell, 38, 44
chemical, 7, 10, 12, 18, 21, 28, 35, 37, 40, 47, 53, 54, 55, 61, 63, 64, 77, 85
chemical composition, 7, 28, 35, 37, 40, 53, 55, 61, 77
chemical etching, 35, 63, 64
chemical stability, 7, 10, 54
chemical vapor deposition, 21
classified, 13, 16
Co, 17, 81, 82, 85, 90
cobalt, 19
Colorado, 90, 91
compatibility, 10
complementary, 29, 33, 57, 77
components, 10, 62, 64
composite, 31
composition, 26, 29, 34, 37, 40, 42, 60, 73, 78
compositions, 26, 32, 77
compounds, 7, 19, 23, 29, 44, 61, 65
computers, 81
concentration, 11, 12, 13, 14, 24, 35, 67
conductive, 34
conductivity, 9, 10, 25
contamination, 31
continuity, 12

controlled, 10, 16, 21, 22, 27
conversion, 65
cooling, 16, 61
coordination, 68
copper, 7, 9, 16, 20, 21, 23, 24, 25, 26, 27, 28, 29, 34, 37, 38, 40, 44, 47, 49, 52, 53, 57, 60, 62, 63, 64, 66, 68, 70, 73, 75, 76, 77, 83, 89, 90, 91
correlation, 15
Coulomb interaction, 16
covalent, 16
covalent bond, 16
covalent bonding, 16
coverage, 21, 78
CRC, 81
critical temperature, 33, 53, 72, 74, 75, 76, 78
cross-sectional, 41, 61
crystal, 15, 16, 23, 24
crystal lattice, 15
crystal structure, 16, 24
crystalline, 18, 34, 35, 57, 60, 62, 66, 69, 72, 73
crystallinity, 35
crystallites, 38, 41, 44, 46, 47, 54, 61, 65, 66, 72
crystallization, 19, 26, 27, 28, 54, 55, 56, 57, 60, 62, 64, 65, 72, 73, 75, 76, 77, 78
CVD, 21
cycles, 22

D

damping, 40
data base, 34
decomposition, 44, 45
defect formation, 75
defects, 15, 16, 18, 24, 76
degradation, 10, 33, 43, 54, 57, 65, 72
degradation mechanism, 43
degradation process, 33
degree, 34
degree of crystallinity, 34
delivery, 67, 69
demand, 7
density, 18, 21, 24, 40

deposition, 10, 18, 20, 21, 22, 23, 24, 25, 26, 27, 31, 38, 40, 47, 57, 61, 62, 78
deposition rate, 22
detection, 8, 32, 53, 62, 63, 64, 75, 77
dielectric, 7, 9, 10, 16, 17, 18, 19, 20, 21, 22, 23, 25, 28, 29, 35, 37, 47, 54, 61, 63
dielectric constant, 9, 25
dielectric materials, 7, 10, 16, 17, 28
dielectrics, 7, 9, 10, 11, 17, 18, 25, 29, 78
diffraction, 8, 16, 34, 38, 39, 44, 46, 51, 61, 67, 75
diffusion, 7, 10, 11, 12, 13, 14, 15, 16, 17, 18, 19, 20, 21, 22, 23, 24, 25, 26, 27, 28, 29, 31, 33, 34, 35, 37, 38, 39, 40, 42, 44, 47, 49, 50, 51, 52, 53, 54, 57, 58, 60, 62, 63, 64, 65, 66, 68, 69, 70, 72, 73, 75, 76, 77, 78, 82, 89
diffusion mechanisms, 15
diffusion process, 13, 15, 16, 19, 26, 27, 29, 33, 50, 52, 53, 57, 62, 70
diffusion rates, 15
diodes, 28
directionality, 21
dislocation, 14
dislocations, 14, 15, 16
distribution, 11, 48, 50, 51, 52, 55, 65, 69
donors, 16
dopant, 16
doping, 85
dry, 10

E

ECD, 20
Einstein, 11
electrical, 7, 9, 10, 17, 18, 21, 22, 25, 37, 42, 44, 53
electrical conductivity, 9, 10
electrical properties, 10, 22, 37, 53
electrochemical, 20
electrochemical deposition, 20
electrodeposition, 20
electromigration, 7, 10, 28, 90
electron, 10, 34, 35, 40, 41, 60, 62
electron density, 34
electron diffraction, 35, 41, 60
electron microscopy, 35
electronic, 28
electronic structure, 28
electrons, 21
electroplating, 20
electrostatic, 11
elongation, 18, 78
emission, 7, 8, 34
energy, 16, 21, 31, 34, 35, 53, 61
energy transfer, 21, 53
English, 85
environmental, 10
environmental influences, 10
equilibrium, 66
etching, 35
Eulerian, 34
evidence, 68
evolution, 49
excess supply, 46
exponential, 15, 16, 25
exposure, 59
extrapolation, 16
eyes, 68

F

failure, 7, 10, 20, 22, 23, 24, 26, 28, 29, 33, 43, 63, 75, 77, 78
family, 20
film, 7, 18, 20, 21, 22, 25, 28, 29, 31, 32, 34, 35, 37, 38, 39, 40, 41, 43, 45, 50, 51, 53, 54, 55, 57, 58, 60, 62, 64, 65, 67, 68, 69, 70, 72, 73, 75, 76, 77, 78
film thickness, 22, 32, 34, 38, 39, 40, 58, 65, 68, 70
films, 7, 15, 18, 20, 21, 22, 23, 24, 25, 26, 27, 28, 31, 35, 41, 45, 46, 51, 53, 54, 70, 72, 75, 76, 77, 78, 89, 91, 94
flow, 22, 26, 31, 40, 44
flow rate, 31, 40
fluctuations, 50
focusing, 34
fracture, 10

G

gas, 21, 22, 24, 26, 33, 44, 49, 58
gases, 22
generation, 70
Germany, 79, 81, 94
Gibbs, 66
Gibbs energy, 66
glass, 25
glow discharge, 7
grain, 10, 13, 14, 15, 16, 18, 19, 24, 25, 27, 28, 29, 37, 38, 46, 50, 53, 57, 58, 65, 68, 69, 70, 75, 76, 78, 82
grain boundaries, 10, 13, 15, 16, 18, 24, 25, 27, 28, 29, 37, 50, 53, 57, 58, 65, 68, 70, 75, 76
grains, 13, 28, 38, 39, 45, 61, 62, 70
graphite, 34, 35
groups, 16, 19
growth, 17, 18, 39, 45, 46, 50, 61, 66, 67, 69, 70, 78
growth rate, 67, 69

H

H_2, 17, 19
heat, 17, 21, 24, 25, 26, 28, 33, 43, 44, 46, 49, 54, 57, 60, 62, 64, 65, 67, 73, 74
heating, 16, 17
heterogeneous, 61
high-frequency, 40
homogeneous, 12, 13, 51, 59, 61
homogenous, 14
hydrazine, 26
hydro, 44
hydrocarbons, 44
hydrofluoric acid, 35
hydrogen, 22, 26

I

identification, 34
images, 35, 59, 62
imaging, 35
implementation, 9, 10, 18, 20, 25, 27, 28, 29
impurities, 61
IMS, 34
in situ, 17
incidence, 17, 33, 38
indication, 57
industrial, 42
industrial application, 42
inequality, 75
infinite, 66
Information Technology, 85
insertion, 24
insulator-semiconductor, 19
integration, 27, 91
integrity, 53, 54, 55, 65, 77
intensity, 38, 39, 44, 46, 49, 57, 62, 65, 68
interaction, 53, 57
interface, 13, 18, 19, 20, 21, 24, 27, 34, 40, 43, 49, 57, 58, 61, 64, 66, 69, 70
interphase, 82
interstitial, 15, 16, 70
interstitials, 15, 16, 70
intrinsic, 12, 18, 24, 27, 28, 41, 43, 66
Investigations, 26
ion bombardment, 24
ion mass spectroscopy, 8, 34, 63
ionization, 21
ions, 7, 10, 16, 17, 21, 31, 34
iron, 19
island, 50
island formation, 50

K

kinetics, 13, 14

L

Langmuir, 87
lattice, 11, 12, 16, 23, 26, 27, 34, 35, 44, 46, 70
lattice parameters, 23, 34, 44
law, 11, 12
lead, 10, 16, 18, 26, 29, 50, 54, 72, 75

leakage, 25
Leibniz, 79
lifetime, 83
linear, 14, 16, 50
literature, 19, 28, 37, 39
London, 81, 82, 85, 92
long-term, 25
low-temperature, 66, 70
LTD, 81

M

magnetron, 21, 31, 77
magnetron sputtering, 31, 77
manufacturing, 33
mathematics, 82
matrix, 70
mechanical, iv, 10, 18, 25, 37, 44
mechanical properties, 10, 25, 44
melt, 61
melting, 9, 16, 19, 23, 26, 66
melting temperature, 16, 19, 23, 26
metal ions, 21
metals, 19, 20, 28
microelectronics, 17
micrometer, 20
microprocessors, 7, 9, 10, 16
microscope, 35
microscopy, 8
microstructure, 7, 10, 11, 18, 23, 24, 26, 29, 33, 35, 37, 40, 41, 43, 49, 51, 53, 54, 60, 63, 66, 67, 68, 70, 72, 73, 77, 78
microwave, 83
migration, 11, 15
mirror, 34
mobility, 18
mole, 12
monochromator, 34
monolayer, 22
morphology, 25, 40
motion, 15
MRS, 81

N

nanocrystalline, 18, 26, 27, 37, 38, 39, 53, 68, 78
nanometers, 32
Nb, 19, 23
network, 9
new, 9, 18, 78
New York, 86, 92, 94
Ni, 34
nickel, 19
nitride, 17, 23, 27, 28, 54, 61, 78
nitrides, 57, 61, 62, 68
nitrogen, 8, 20, 22, 26, 28, 34, 46, 55, 61, 77
nodes, 22
nucleation, 20, 22, 24, 54, 61, 66, 69, 70, 73, 75, 76, 78
nucleation layer, 20

O

observations, 49
optical, 7, 8, 34
optimization, 27
organ, 21
organic, 22
organic compounds, 22
organometallic, 21
orientation, 41, 44, 47
orthorhombic, 45, 54
oscillations, 40, 49, 57, 58
oxidation, 17, 28, 59
oxide, 7, 9, 10, 17, 22, 23, 24, 25, 31, 37, 40, 44, 45, 47, 52, 54, 55, 57, 59, 61, 62, 64, 67, 73, 75, 77
oxide thickness, 45
oxygen, 17, 22, 24, 31, 34, 44, 53, 54, 57, 62

P

PACS, 8
pairing, 16
parameter, 23, 26, 46
particles, 12

Index

passivation, 16, 57, 64
performance, 7, 9, 21, 25, 33, 37
periodic, 34
periodic table, 34
permittivity, 9, 25
phase diagram, 61, 66, 86, 93
phase transformation, 68
photoelectron spectroscopy, 23
photons, 34
physical properties, 9, 23, 27, 53
planar, 66
plasma, 21, 22, 26, 31, 34
point defects, 15, 16
polycrystalline, 18, 22, 24, 25, 38, 40, 41, 44, 49, 51, 65, 75, 92
polymer, 19
poor, 42
porous, 22, 25, 78
powder, 26, 34
power, 21, 31, 35
preparation, 22, 31, 35, 64
pressure, 22, 24, 31, 33
probe, 7, 42
procedures, 92
promote, 22, 28
propagation, 9, 25
property, 7, 29

R

radiation, 17, 34
radio, 21, 31, 34
radius, 14
range, 7, 16, 24, 27
reactant, 22
reaction temperature, 16, 55, 66, 70
redistribution, 52, 53, 72, 77, 78
reduction, 25, 27
reflection, 38, 40, 41, 83
refractory, 19, 23, 28
regular, 70
relationship, 14
relationships, 16
reliability, 81
research, 23, 27

reservoirs, 66
resistance, 7, 9, 10, 28, 90
resistivity, 7, 9, 20, 21, 23, 26, 27, 37, 39, 42, 53, 78
resolution, 34, 35
retardation, 26
roadmap, 10
room temperature, 17, 23, 31
roughness, 49, 57, 58
ruthenium, 20, 22
Rutherford, 32

S

sample, 16, 17, 25, 33, 34, 35, 38, 39, 45, 46, 48, 49, 50, 51, 52, 53, 55, 56, 57, 58, 59, 60, 62, 63, 64, 66, 67, 68, 69, 70, 71, 72, 73, 74, 75, 76
sample design, 62
scaling, 7, 9, 10, 18, 20, 29, 78
scanning electron microscopy, 8, 49
scattering, 7, 10, 33, 41, 49, 77
seed, 20, 21
seed layer, 20
segregation, 16, 18
SEM, 49, 50
sensitivity, 32
services, iv
shape, 50
signals, 9
signs, 39, 41, 46, 53, 68, 74
silicon, 7, 9, 10, 16, 17, 20, 23, 24, 25, 26, 28, 31, 34, 37, 38, 52, 54, 57, 61, 62, 63, 64, 66, 67, 68, 69, 70, 72, 73, 75, 77, 78, 83
silicon dioxide, 83
single crystals, 16
SiO_2, 18, 19, 20, 21, 23, 25, 26, 27, 28, 29, 32, 33, 35, 38, 39, 41, 43, 45, 46, 47, 48, 49, 50, 52, 54, 55, 56, 57, 58, 60, 61, 62, 63, 64, 65, 66, 67, 68, 70, 72, 73, 75, 76, 77
sites, 16, 62, 70
software, 34
solid state, 62, 67
solubility, 16, 17, 20, 23, 62
species, 14, 60

spectroscopic methods, 77
spectroscopy, 7, 8, 34, 35, 40
spin, 25
sputtering, 21, 31, 90
stability, 7, 10, 18, 20, 24, 26, 28, 51, 53, 76, 89
stabilization, 45
standards, 92
stiffness, 10
stoichiometry, 55, 61
strains, 34
strength, 17, 25, 70
stress, 10, 11, 24, 27, 28, 37, 38, 41, 43, 45, 47, 66
substrates, 27, 29, 31, 78
superposition, 39
supply, 44, 46
surface area, 50
surface diffusion, 15
surface energy, 50
surface layer, 14, 57
surface region, 59
surface roughness, 34, 40
systematic, 28, 33, 63
systems, 12, 33, 47, 52, 54, 56, 71, 73, 78

T

tantalum, 8, 22, 23, 24, 28, 34, 38, 44, 47, 53, 66, 68, 90
technological, 21
technology, 10, 21, 22, 27
TEM, 35, 39, 41, 45, 47, 57, 59, 60, 62, 63, 65
temperature, 9, 10, 11, 15, 16, 19, 23, 24, 26, 32, 43, 46, 60, 66, 69, 73, 75, 76
temperature dependence, 15
temperature gradient, 11
TEOS, 31
theoretical, 45, 66
thermal, 7, 9, 10, 15, 17, 18, 20, 22, 23, 24, 25, 26, 27, 28, 29, 33, 35, 37, 40, 43, 44, 45, 46, 47, 54, 55, 57, 61, 63, 70, 72, 75, 76, 77, 78
thermal activation, 46
thermal expansion, 37

thermal stability, 7, 17, 20, 23, 24, 25, 26, 27, 28, 29, 43, 44, 47, 54, 57, 61, 72, 75, 77, 78
thermal treatment, 35, 63, 70, 76, 77
thermodynamic, 12
thermodynamics, 66
thin film, 13, 22, 91
thin films, 22, 91
three-dimensional, 9, 18
Ti, 19, 21, 23
timing, 22
toughness, 10
transfer, 70
transformation, 23, 26, 45, 46, 61, 68, 72
transistor, 10
transistors, 9
transition, 66
transition metal, 66
transmission, 7, 8, 35
transmission electron microscopy, 7, 8, 92
transport, 16, 82
trend, 60
tungsten, 21
two-dimensional, 16

U

uniform, 22
USSR, 92

V

vacancies, 15
vacuum, 20, 24, 32, 33, 34, 40, 44
vapor, 20, 21, 22
velocity, 11
visible, 49, 68, 72, 74

W

wet, 35, 63, 64
wettability, 20

X

X-ray, 7, 8, 16, 17, 23, 26, 33, 35, 39, 41, 54, 72, 77, 78, 92

X-ray absorption, 39

X-ray diffraction (XRD), 7, 8, 16, 26, 33, 34, 38, 39, 41, 43, 44, 45, 48, 49, 51, 54, 56, 57, 60, 62, 63, 66, 68, 69, 71, 72, 73, 74, 75, 78, 92

X-ray photoelectron spectroscopy (XPS), 41

X-ray reflectometry, 7, 33

Z

Zn, 26